SORBENTS FOR LIQUID HAZARDOUS SUBSTANCE CLEANUP AND CONTROL

SORBENTS FOR LIQUID HAZARDOUS SUBSTANCE CLEANUP AND CONTROL

by

Robert W. Melvold and Steven C. Gibson

Environmental Monitoring and Services, Inc.
Combustion Engineering
Newbury Park, California

Robert Scarberry

JRB Associates
McLean, Virginia

NOYES DATA CORPORATION
Park Ridge, New Jersey, U.S.A.

Library of Congress Catalog Card Number 87-31550
ISBN: 0-8155-1159-0
ISSN: 0090-516X
Printed in the United States

Published in the United States of America by
Noyes Data Corporation
Mill Road, Park Ridge, New Jersey 07656

10 9 8 7 6 5 4 3 2 1

Library of Congress Cataloging-in-Publication Data

Melvold, Robert W.
Sorbents for liquid hazardous substance cleanup and control / by Robert W. Melvold and Steven C. Gibson and Robert Scarberry.
p. cm. -- (Pollution technology review, ISSN 0090-516X ; no. 150)
Bibliography: p.
Includes index.
ISBN (invalid) 0-8155-1159-0
1. Hazardous substances--Handbooks, manuals, etc. 2. Sorbents--Handbooks, manuals, etc. I. Gibson, Steven C. II. Scarberry, Robert. III. Title. IV. Series
T55.3.H3M43 1988
604.7--dc19 87-31550
CIP

Foreword

This book focuses on the selection and use of sorbents for cleanup or control of liquids designated as hazardous substances by CERCLA (Comprehensive Environmental Response, Compensation, and Liability Act–1980). It contains information and data which will be useful to personnel in federal, state, and local governments, as well as the private sector, who must select, purchase, stockpile, apply, collect, or dispose of sorbents for the purpose of controlling hazardous substance releases.

The evaluation and selection of a sorbent that is effective and, when cost is a factor, economical, can be a difficult task. One source of difficulty is that many factors influence the performance and cost of using sorbents for spill cleanup operations. Factors that require consideration include safety, sorption capacity, desorption characteristics, sorbent cost, availability, and ease of application, collection, regeneration, and disposal. The difficulties in evaluating these factors and judging their relative effects on overall performance are compounded by the lack of information, data, and guidance on sorbent use for liquid hazardous substance control, and the dispersion and nonuniformity of much of the data that are available.

This manual is an attempt to consolidate available information, literature data, and data generated in the laboratory, and to place the combined findings in a format that is readily accessible to the user.

To utilize the manual, the user must first identify the spilled liquid. If it is one of the 212 liquid hazardous substances addressed, a reference is provided to one of 26 "Sorbent Selection and Use Guides." Each of the Guides enables the user to rapidly identify generic sorbent classes, physical forms, and methods for application and collection that are most suitable for each of three different chemical release-control scenarios: (1) a spill onto land, (2) a floating release into water, and (3) immobilization for landfilling.

In addition, the manual contains "Sorbent Data Sheets" for 12 generic classes of sorbents other than activated carbon. These data sheets contain information on manufacturers, acquisition costs, sorbent density and sorbent limitations. The sorption capacities of 190 sorbent-chemical pairs were determined and recorded. The manual also includes cost estimation procedures, test methods, hazardous liquid physical properties, the rationale for the release-control scenarios, and a detailed description of the sorbent selection rationale.

The information in the book is from *A Guidance Manual for Selection and Use of Sorbents for Liquid Hazardous Substance Cleanup and Control,* prepared by Robert W. Melvold and

Steven C. Gibson of Environmental Monitoring and Services, Inc., a subsidiary of Combustion Engineering, and Robert Scarberry of JRB Associates, for the U.S. Environmental Protection Agency, July 1987.

The table of contents is organized in such a way as to serve as a subject index and provides easy access to the information contained in the book.

Advanced composition and production methods developed by Noyes Data Corporation are employed to bring this durably bound book to you in a minimum of time. Special techniques are used to close the gap between "manuscript" and "completed book." In order to keep the price of the book to a reasonable level, it has been partially reproduced by photo-offset directly from the original report and the cost saving passed on to the reader. Due to this method of publishing, certain portions of the book may be less legible than desired.

ACKNOWLEDGMENT

The authors gratefully acknowledge the guidance and support provided by Michael Royer of the Releases Control Branch (RCB), Hazardous Waste Engineering Research Laboratory (HWERL), EPA. The authors also express their gratitude to the following individuals who provided technical assistance in the development of the manual: Dr. Gary Bennett, the University of Toledo, College of Engineering; Dr. Walter Unterberg, Rockwell International; Ms. Patricia Scofield (on educational leave from Rockwell International), Mr. Fitzhugh Bush III, EMSI, and Dr. William Ellis, JRB Associates.

In addition, the authors thank the following persons for their comments and guidance during preparation of the manual: Mr. Leo T. McCarthy, Jr., Mr. John S. Farlow, and Mr. Frank J. Freestone, EPA–RCB, HWERL; Mr. Paul Cassidy, EPA–Office of Solid Waste; Mr. Don Berger, EPA–Region I; Mr. Mark Hall, Technical Assistance Team, Region I; Mr. Paul Elliot (retired)–EPA Region II; Mr. David Lopez, EPA–Region VI; Mr. William Keffer and Mr. Gary Snodgrass, EPA–Region VII; Mr. Andre Zownir and Mr. Robert Cibulskis, EPA–Environmental Response Team; Mr. Jack Sinclair, United States Coast Guard (USCG)–Washington, DC; CWO Rushnak, J.C. Card, and W. Isele, USCG Safety Office, St. Louis, MO; Petty Officer Kenyon, USCG Water Pollution Response Office, New York, NY; and Brett Betts, Department of Ecology–State of Washington, Olympia, WA.

NOTICE

The materials in this book were prepared as accounts of work sponsored by the U.S. Environmental Protection Agency. It has been approved for publication as an EPA document. On this basis the Publisher assumes no responsibility nor liability for errors or any consequences arising from the use of the information contained herein. Mention of trade names or commercial products does not constitute endorsement or recommendation for use by the Agency or the Publisher.

Final determination of the suitability of any information or procedure for use contemplated by any user, and the manner of that use, is the sole responsibility of the user. The reader is warned that caution must always be exercised when dealing with hazardous materials, and expert advice should be sought at all times.

Contents and Subject Index

1. Sorbent Selection and Use

INTRODUCTION

This Manual is for persons who make decisions regarding the selection, acquisition, application, collection, reuse, or disposal of sorbents for the purpose of controlling releases of liquid hazardous substances or landfilling of liquid hazardous substances. A primary target audience for the Manual is EPA and other Federal On-Scene Coordinators, their technical support personnel, and their State Counterparts. The manual also provides useful information and data to industry and spill cleanup personnel who respond to releases releases of hazardous substances or who must immobilize liquid hazardous substances prior to landfilling.

Sorbents, when properly selected and used, are a versatile tool for the On-Scene Coordinator (OSC) or other spill cleanup manager. Sorbents are used to (1) soak up and thereby retard the movement of liquid chemical pools on land, (2) reduce spreading or evaporation of hazardous substances that float on water, and (3) solidify liquid spill residuals to make them acceptable for landfilling when other options are unacceptable for technical, economic, or regulatory reasons.

The evaluation and selection of a sorbent that is effective and, when cost is a factor, economical, can be a difficult task. One source of difficulty is that many factors influence the performance and cost of using sorbents for spill cleanup operations. Factors that require consideration include safety, sorption capacity, desorption characteristics, sorbent cost, availability, and ease of application, collection, regeneration, and disposal. The difficulties in evaluating these factors and judging their relative effect on overall performance are compounded by the lack of information, data, and guidance on sorbent use for liquid hazardous substance control, and the dispersion and non-uniformity of much of the data that are available.

This Manual is an attempt to consolidate available information, literature data, and data generated in the laboratory by the authors, and to place the combined findings in a format that is readily accessible to the user. The Manual primarily addresses chemicals that are (1) neat, (2) liquid at ambient conditions, and (3) designated as hazardous by CERCLA Section 101(14). The spill scenarios that are addressed are limited to (1) releases of less than 5000 gallons and (2) gross cleanup operations, as compared to operations that produce "polished" effluent streams. As a consequence of excluding "polishing" operations, activated carbon, a widely used sorbent for spill cleanup, is not addressed in this Manual. In addition, the sorbents that are addressed in this Manual are limited to those that are readily available in at least some, although not necessarily all, regions of the country.

HOW TO USE THE MANUAL

The following steps should be followed in sequence for the most effective use of the Manual:

1. Become familiar with the general contents and layout of the Manual.

2. Once a spill or release of a hazardous liquid has occurred or is likely, proceed to Table 1-1 or Table 1-2 (the CERCLA Liquids Indices) to identify the "Sorbent Selection and Use Guide" (Guide) for the chemical class of the hazardous liquid of interest. The hazardous liquid names are in alphabetical order in Table 1-1 and in numerical order (DOT ID number) in Table 1-2, followed by the Guide numbers. The Chemical Abstract Services (CAS) number is also provided for each chemical in Tables 1-1 and 1-2. The indexing tables also include a listing of the hazards involved upon release of a particular liquid and the behavior of the liquid in water.

   NOTE: This Manual indicates some, but not necessarily all, the known hazards of the identified liquids, and the user is expected to make an independent assessment of the hazards and the appropriate safety measures.

3. Consult the specified Guide (the Guides are located on pp 1-32 to 1-58). Each Guide lists, in recommended order of consideratin for use, as many as six sorbents for each applicable release scenario. In order to determine whether the rationale employed to prioritize the sorbents in the Guide is reasonable for the situation facing the user, consult Section 6, "Spill Scenario Rationale," and Section 9, "Sorbent Selection Rationale." The reader may also want to employ the rationales described for sorbents not addressed by the Manual.

4. Determine whether any of the noted "Limitations" precludes the use of the first listed sorbent or sorbents. If so, proceed down the ordered list of sorbents until one is located that is not limited by site conditions or logistic requirements. Note that sorbents preceded by the same prioritization number have been rated equivalent and are perceived as having the same general potential for performing effectively. Also, the user should keep clearly in mind that the sorbent rankings, except for the Landfill scenario, were developed without consideration of cost.

5. After identifying suitable sorbent classes and forms, consult Section 2, "Sorbent Data," which contains information on sorbent producers, sorption capacity (by sorbent class and chemical class only) and hazardous liquid/water preference indices. Compatibility information is included with the sorption capacity data.

6. After selecting a sorbent, use the optimum application and collection methods listed in the Guide or choose other methods more appropriate for

the situation. (Other methods may be preferable if conditions at your site are atypical and not used in the evaluation.) Consult Section 3 for guidance on equipment and techniques for application, collection, and disposal of sorbents.

7. If cost is an important consideration, the user should examine Section 4, "Cost Estimation Procedures and Data", which contains cost data on equipment, labor, transportation, disposal, and sorbents.

8. Refer to the remaining sections of the Manual as needed:

   a. Section 5 describes sorbent test methods.

   b. Section 7 lists sources of information.

   c. Section 8 contains data on the CERCLA liquids' specific gravity, solubility, surface tension, and viscosity.

EXAMPLE OF USING THE MANUAL — 100 GALLON ETHYLBENZENE SPILL ON LAND

The following case illustrates the process a manual user should follow when a spill or release occurs. In this example, 100 gallons[1] of ethylbenzene has been released on a windy day and flowed into a depression covered with dense foliage. The reader would first refer to ethylbenzene in the indexing table and find that it belongs to a chemical class "Aromatic Hydrocarbons", and that sorbent selection and use information is in Guide No. 10. He will note that ethylbenzene is a flammable, insoluble floater. After referring to Guide No. 10, the user would look under "Landspill-Small" for the ordered list of sorbents and would then scan the "Limitations" column for conditions that applied to his particular case (the demarcation between small and medium landspills has been set at 500 gallons). The user would then eliminate "(1) Feathers/pl" because of the dense ground cover, "(2) CLP/p" because of the wind, and "(2) CLP/pl," also because of the dense ground cover. Sorbent clay and treated clay/treated natural organic mix, both particulate sorbents, would remain as candidates for use. Though the optimum method for both application and collection is listed as "shovel," a mechanized method of collection may be more suitable here because of the dense ground cover present. If the user wants more information on the sorbents or on application and collection methods, he should refer to the Sorbent Data Sheets in Section 2, and/or he should review Section 3 on logistics. If the same spill or release had occurred in a small pond or large lake, the user might have selected cross-linked polymer in the pillow form, subject to its timely availability.

-----

1 To convert English units to metric (Sl units), use Table of Conversions on page 154.

CERCLA LIQUIDS INDICES AND SORBENT SELECTION AND USE GUIDES

The remainder of Section 1 contains the CERLCA Liquids Name Index (Table 1-1) and Numerical Index (Table 1-2) and 26 Sorbent Selection and Use Guides (Guides). The CERCLA liquids indices list CERCLA-designated hazardous liquids or their DOT ID numbers, the Guide number and the chemical class (to which each liquid has been assigned), the Chemical Abstract Service (CAS) number (for the liquid), the hazards, other than toxicity (a hazard posed by most CERCLA substances), and the behavior of the liquid in water. A total of 212 CERCLA-designated hazardous liquids were identified. A liquid is defined as a substance that possesses a melting point at or below 77°F (25°C) and a boiling point at or above 50°F (10°C).

The Guides present information on the use of sorbents for one to three applicable major scenarios, i.e., Landspill, Floating Spill, and Landfill. The 212 CERCLA liquids are categorized into 27 chemical classes that are descriptive of the chemical reactivity of the CERCLA liquids. Since sorption capacity data are required to prioritize sorbents for consideration and because it was not practical to experimentally evaluate all of the CERCLA hazardous liquids, representative hazardous liquids therefore were selected from each chemical class. Selection of a representative hazardous liquid was based on similarities in aqueous solubility, specific gravity, liquid surface tension, and viscosity, where these data were available, with other liquids in the same class. Sorption capacities were then measured experimentally and reported as representative of a chemical class (see Table 2-1). A list of the chemical classes and Sorbent Selection and Use Guide numbers to which each hazardous liquid was assigned is given as follows:

Chemical Class	Guide#	Chemical Class	Guide#
Acidic compounds, inorganic	1	Esters and ethers	15
Acidic compounds, organic	2	Ethers, halogenated	16
Alcohols and glycols	3	Halides, inorganic	17
Aldehydes	4	Heavy metals	18
Aliphatic hydrocarbons	5	Hydrazines and hydrazides	19
Aliphatics, halogenated	6	Ketones	20
Amides, anilides, and imides	7	Nitro/nitroso compounds	21
Amines, alkyl	8	Organophosphates	22
Amines, aryl	9	Oxides, alkylene	23
Aromatic hydrocarbons	10	Peroxides	24
Aromatics, halogenated	11	Phenols and cresols	25
Basic compounds	12	Sulfates and sulfites	26
Cyanates and isocyanates	13	Sulfides and mercaptans	27
Cyanides and nitriles	14		

It was originally planned to have 27 Guides (a Guide for each chemical class), but only 26 Guides have been produced, due to an insufficiency of data for Guide No. 7 (Amides, anilides, and imides). Each of the completed Guides contains several prioritized lists of generic sorbents for consideration relative to each of the three scenarios generally applicable to a release of the particular class of CERCLA liquids that the Guide describes. For instance, if a CERCLA liquid class, as a whole, sinks and is only slightly soluble in water (less than 1%), information relative to the Landspill and Landfill Scenarios would be found in that Guide, but information on "floaters" would not be included. The prioritized sorbents are cross-referenced to the Sorbent Data Sheets of Section 2 by page number, which is included in each Guide just after the name and type of sorbent.

Application and collection procedures are listed in the Guide and are based on assessment and selection of the procedures identified as optimum for use with each listed generic sorbent for the scenario involved. These procedures do not consider site-specific limitations which may affect application and collection. For the Landspill and Floating Spill Scenarios, limitations that would prevent use of a class or type of sorbent are also given. The Floating Spill Scenario is further subdivided to provide the above information for spills into non-flowing and flowing water bodies. Cost, as a function of sorption capacity, is used to provide the basis for comparison of sorbents for the Landfill Scenario since this scenario does not involve an emergency situation.

Sorbent Prioritization tables are used in conjunction with compatibility and wettability data to develop the prioritized lists of sorbents included in the Guides. The Sorbent ranking process is described in Section 9, while incompatible and nonwetting sorbent-chemical combinations are indicated in the Sorption Capacity table in Section 2 (compatibility and wettability is implied if a sorption capacity value appears in the table).

TABLE 1-1. CERCLA LIQUIDS NAME INDEX

Hazardous Substance	Chemical Class	Guide No.	CAS No.*	Hazard(s), in Addition to Toxicity	Behavior in Water
Acetaldehyde	Aldehydes	4	75-07-0	Flammable Polymerizable	Soluble Floater
Acetic acid	Acidic compounds, organic	2	64-19-7	Combustible Corrosive	Soluble Sinker
Acetic anhydride	Acidic compounds, organic	2	108-24-7	Combustible Corrosive	Soluble Sinker
Acetone	Ketones	20	67-64-1	Flammable	Soluble Floater
Acetone cyanohydrin	Cyanides and nitriles	14	75-86-5	Combustible w/toxic products Poison	Soluble Floater
Acetonitrile	Cyanides and nitriles	14	75-05-8	Flammable w/toxic products	Soluble Floater
Acetophenone	Ketones	20	98-86-2	Combustible	Insoluble Sinker
Acetyl bromide	Aliphatics, halogenated	6	506-96-7	Flammable w/toxic products Corrosive Reactive	Decomposes (Sinker)
Acetyl chloride	Aliphatics, halogenated	6	75-36-5	Flammable w/toxic products Corrosive Reactive	Decomposes (Sinker)
Acrolein	Aldehydes	4	107-02-8	Flammable Polymerizable Poison	Soluble Floater
Acrylic acid	Acidic compounds, organic	2	79-10-7	Combustible Corrosive Polymerizable	Soluble Sinker
Acrylonitrile	Cyanides and nitriles	14	107-13-1	Flammable w/toxic products Polymerizable Potential carcinogen Poison	Soluble Floater
Allyl alcohol	Alcohols and glycols	3	107-18-6	Flammable Poison	Soluble Floater

*Chemical Abstract Services Number

(continued)

TABLE 1-1. (continued)

Hazardous Substance	Chemical Class	Guide No.	CAS No.	Hazard(s), in Addition to Toxicity	Behavior in Water
Allyl chloride	Aliphatics, halogenated	6	107-05-1	Flammable w/toxic products	Insoluble Floater
Ammonium hydroxide	Basic compounds	12	1336-21-6	Corrosive	Soluble Floater
Ammonium sulfide (in aqueous solution)	Sulfides and mercaptans	27	12135-76-1	Flammable w/toxic products	Soluble Sinker
Amyl acetate	Esters and ethers	15	628-63-7	Flammable	Insoluble Floater
Aniline	Amines, aryl	9	62-53-3	Combustible w/toxic products Poison	Soluble Sinker
Antimony pentachloride	Halides, inorganic	17	7647-18-9	Corrosive Reactive	Decomposes (Sinker)
Arsenic trichloride	Halides, inorganic	17	7784-34-1	Corrosive Reactive Poison	Decomposes (Sinker)
Benzal chloride	Aromatics, halogenated	11	98-87-3		Insoluble Sinker
Benzene	Aromatic hydrocarbons	10	71-43-2	Flammable Potential carcinogen	Insoluble Floater
Benzenesulfonyl chloride	Acidic compounds, organic	2	98-09-9	Combustible w/toxic products	Insoluble Sinker
Benzonitrile	Cyanides and nitriles	14	100-47-0	Combustible w/toxic products	Soluble Sinker
Benzotrichloride	Aromatics, halogenated	11	98-07-7	Combustible w/toxic products Corrosive	Decomposes (Sinker)
Benzoyl chloride	Aromatics, halogenated	11	98-88-4	Combustible w/toxic products Corrosive	Decomposes (Sinker)
Benzyl chloride	Aromatics, halogenated	11	100-44-7	Combustible w/toxic products Corrosive Reactive	Insoluble Sinker
2,2′-Bioxirane	Oxides, alkylene	23	1464-53-5	Potential carcinogen	Decomposes (Sinker)

(continued)

TABLE 1-1. (continued)

Hazardous Substance	Chemical Class	Guide No.	CAS No.	Hazard(s), in Addition to Toxicity	Behavior in Water
Bis(2-chloroethoxy) methane	Ethers, halogenated	16	111-91-1		Soluble Sinker
Bis(2-chloroethyl) ether	Ethers, halogenated	16	111-44-4	Combustible w/toxic products Poison	Soluble Sinker
Bis(2-chloroisopropyl)-ether	Ethers, halogenated	16	108-60-1	Combustible w/toxic products	Insoluble Sinker
Bis(chloromethyl) ether	Ethers, halogenated	16	542-88-1	Combustible w/toxic products Potential carcinogen	Soluble Sinker
Bis(2-ethylhexyl) phthalate	Esters and ethers	15	117-81-7		Insoluble Floater
Bromoacetone (Note 2)	Ketones	20	598-31-2	Poison	Soluble Sinker
Bromoform	Aliphatics, halogenated	6	75-25-2		Insoluble Sinker
4-Bromophenyl phenyl ether	Ethers, halogenated	16	101-55-3	Combustible w/toxic products	Insoluble Sinker
1-Butanol	Alcohols and glycols	3	71-36-3	Flammable	Soluble Floater
2-Butanone peroxide (Note 2)	Peroxides	24	1338-23-4	Explosive Oxidizer Combustible	Soluble Sinker
Butyl acetate	Esters and ethers	15	123-86-4	Flammable	Soluble Floater
Butyl benzyl phthalate	Esters and ethers	15	85-68-7		Insoluble Sinker
Butylamine	Amines, alkyl	8	109-73-9	Flammable w/toxic products	Soluble Floater
Butyric acid	Acidic compounds, organic	2	107-92-6	Combustible	Soluble Floater
Carbon disulfide	Sulfides and mercaptans	27	75-15-0	Flammable w/toxic products	Insoluble Sinker
Carbon tetrachloride	Aliphatics, halogenated	6	56-23-5	Potential carcinogen	Insoluble Sinker

(continued)

TABLE 1-1. (continued)

Hazardous Substance	Chemical Class	Guide No.	CAS No.	Hazard(s), in Addition to Toxicity	Behavior in Water
Chloral	Aldehydes	4	75-87-6	Combustible w/toxic products Corrosive Potential carcinogen	Soluble Sinker
Chloroacetaldehyde	Aldehydes	4	107-20-0	Combustible w/toxic products Polymerizable	Insoluble Sinker
Chlorobenzene	Aromatics, halogenated	11	108-90-7	Flammable w/toxic products	Insoluble Sinker
Chlorodibromomethane	Aliphatics, halogenated	6	124-48-1		Insoluble Sinker
Chloroethane	Aliphatics, halogenated	6	75-00-3	Flammable w/toxic products	Insoluble Floater
2-Chloroethyl vinyl ether	Ethers, halogenated	16	110-75-8	Flammable w/toxic products	Soluble Sinker
Chloroform	Aliphatics, halogenated	6	67-66-3	Potential carcinogen	Insoluble Sinker
Chloromethyl methyl ether	Ethers, halogenated	16	107-30-2	Flammable w/toxic products Potential carcinogen Poison	Insoluble Sinker
2-Chlorophenol	Phenols and cresols	25	95-57-8	Combustible w/toxic products	Soluble Sinker
4-Chlorophenyl phenyl ether	Ethers, halogenated	16	7005-72-3	Combustible w/toxic products	Insoluble Sinker
3-Chloropropionitrile	Cyanides and nitriles	14	542-76-7	Combustible w/toxic products	Soluble Sinker
Chlorosulfonic acid	Acidic compounds, organic	2	7790-94-5	Corrosive Reactive	Soluble Sinker
Creosote	Phenols and cresols	25	8001-58-9	Combustible Potential carcinogen	Insoluble Sinker
Cresol	Phenols and cresols	25	1319-77-3	Combustible	Soluble Sinker
Crotonaldehyde	Aldehydes	4	4170-30-3 123-73-9	Flammable	Soluble Floater

(continued)

TABLE 1-1. (continued)

Hazardous Substance	Chemical Class	Guide No.	CAS No.	Hazard(s), in Addition to Toxicity	Behavior in Water
Cumene	Aromatic hydrocarbons	10	98-82-8	Combustible	Insoluble Floater
Cyanogen chloride	Cyanides and nitriles	14	506-77-4	Poison	Soluble Sinker
Cyclohexane	Aliphatic hydrocarbons	5	110-82-7	Flammable	Insoluble Floater
Cyclohexanone	Ketones	20	108-94-1	Combustible	Soluble Floater
2,4-D esters	Esters and ethers	15	94-11-1	Combustible w/toxic products	Insoluble Sinker
Diazinon	Organophosphates	22	333-41-5	Combustible w/toxic products	Insoluble Sinker
1,2-Dibromo-3-chloropropane (Note 2)	Aliphatics, halogenated	6	96-12-8	Combustible w/toxic products Potential carcinogen	Insoluble Sinker
Di-n-butylphthalate	Esters and ethers	15	84-74-2		Insoluble Sinker
Dichlorobenzene (all isomers)	Aromatics, halogenated	11	25321-22-6	Combustible w/toxic products	Insoluble Sinker
o-Dichlorobenzene	Aromatics, halogenated	11	95-50-1	Combustible w/toxic products	Insoluble Sinker
m-Dichlorobenzene	Aromatics, halogenated	11	541-73-1	Combustible w/toxic products	Insoluble Sinker
Dichlorobromomethane	Aliphatics, halogenated	6	75-27-4	Combustible w/toxic products	Insoluble Sinker
1,4-Dichloro-2-butene	Aliphatics, halogenated	6	764-41-0	Flammable w/toxic products	Insoluble Sinker
1,1-Dichloroethane	Aliphatics, halogenated	6	75-34-3	Flammable w/toxic products	Insoluble Sinker
1,2-trans-Dichloroethylene	Aliphatics, halogenated	6	156-60-5	Flammable w/toxic products	Insoluble Sinker
Dichloropropane (all isomers)	Aliphatics, halogenated	6	26638-19-7	Flammable w/toxic products	Insoluble Sinker
Dichloropropene (all isomers)	Aliphatics, halogenated	6	26952-23-8	Flammable w/toxic products	Insoluble Sinker

(continued)

TABLE 1-1. (continued)

Hazardous Substance	Chemical Class	Guide No.	CAS No.	Hazard(s), in Addition to Toxicity	Behavior in Water
1,3-Dichloropropene	Aliphatics, halogenated	6	542-75-6	Flammable w/toxic products	Insoluble Sinker
Dichloropropene-dichloro-propane mixture	Aliphatics, halogenated	6	8003-19-8	Flammable w/toxic products	Insoluble Sinker
2,2-Dichloropropionic acid	Acidic compounds, organic	2	75-99-0	Corrosive	Soluble Sinker
Dichlorvos	Organophosphates	22	62-73-7	Poison	Insoluble Sinker
Diethylamine	Amines,alkyl	8	109-89-7	Flammable w/toxic products Corrosive	Soluble Floater
Diethylarsine (Note 2)	Heavy metals	18	692-42-2	Flammable w/toxic products	Insoluble Sinker
N,N´-Diethylhydrazine	Hydrazines and hydrazides	19	1615-80-1	Combustible w/toxic products Potential carcinogen	Soluble Floater
O,O-Diethyl-S-methyl dithiophosphate (Note 2)	Organophosphates	22	3288-58-2		Insoluble Sinker
Diethyl-p-nitrophenyl phosphate	Organophosphates	22	311-45-5		Insoluble Sinker
Diethyl phthalate	Esters and ethers	15	84-66-2		Insoluble Sinker
O,O-Diethyl-O-pyrazinyl phosphorothioate	Organophosphates	22	297-97-2		Insoluble Sinker
Dihydrosafrole (Note 2)	Esters and ethers	16	94-58-6	Potential carcinogen	Insoluble Sinker
Diisopropyl fluoro-phosphate	Organophosphates	22	55-91-4	Corrosive	Soluble Sinker
alpha,alpha-Dimethylben-zylhydroperoxide	Peroxides	24	80-15-9	Explosive Combustible	Insoluble Sinker

(continued)

TABLE 1-1. (continued)

Hazardous Substance	Chemical Class	Guide No.	CAS No.	Hazard(s), in Addition to Toxicity	Behavior in Water
Dimethyl carbamoyl chloride (Note 2)	Acidic compounds, organic	2	79-44-7	Combustible w/toxic products Reactive Potential carcinogen	Soluble Sinker
alpha,alpha-Dimethyl-phenethylamine (Note 1)	Amines, alkyl	8	122-09-8		Soluble Floater
Dimethyl phthalate	Esters and ethers	15	131-11-3	Combustible	Insoluble Sinker
Dimethyl sulfate	Sulfates and sulfites	26	77-78-1	Combustible w/toxic products Corrosive Potential carcinogen	Soluble Sinker
1,1-Dimethylhydrazine	Hydrazines and hydrazides	19	57-14-7	Flammable w/toxic products Corrosive Potential carcinogen Poison	Soluble Floater
1,2-Dimethylhydrazine	Hydrazines and hydrazides	19	540-73-8	Flammable w/toxic products Corrosive Potential carcinogen	Soluble Floater
Di-n-octylphthalate	Esters and ethers	15	117-84-0		Insoluble Floater
1,4-Dioxane	Esters and ethers	15	123-91-1	Flammable Potential carcinogen	Soluble Sinker
Dipropylamine	Amines, alkyl	8	142-84-7	Flammable w/toxic products	Soluble Floater
Dodecylbenzenesulfonic acid	Acidic compounds, organic	2	27176-87-0	Combustible w/toxic products	Soluble Sinker
Endrin aldehyde (Note 2)	Aldehydes	4	930-55-2		Insoluble Sinker
Epichlorohydrin	Oxides, alkylene	23	106-89-8	Flammable w/toxic products	Insoluble Sinker
Ethion	Organophosphates	22	563-12-2	Poison	Insoluble Sinker
Ethyl acetate	Esters and ethers	15	141-78-6	Flammable	Soluble Floater

(continued)

TABLE 1-1. (continued)

Hazardous Substance	Chemical Class	Guide No.	CAS No.	Hazard(s), in Addition to Toxicity	Behavior in Water
Ethyl acrylate	Esters and ethers	15	140-88-5	Flammable Polymerizable	Soluble Floater
Ethyl cyanide	Cyanides and nitriles	14	107-12-0	Flammable w/toxic products	Soluble Floater
Ethyl-4,4′-dichlorobenzilate (Note 1)	Esters and ethers	15	510-15-6		Insoluble Sinker
Ethyl ether	Ethers and ethers	15	60-29-7	Explosive, upon standing Flammable	Soluble Floater
Ethyl methacrylate	Esters and ethers	15	97-63-2	Flammable Polymerizable	Insoluble Floater
Ethyl methanesulfonate (Note 2)	Esters and ethers	15	62-50-0		Soluble Sinker
Ethylbenzene	Aromatic hydrocarbons	10	100-41-4	Flammable	Insoluble Floater
Ethylene dibromide	Aliphatics, halogenated	6	106-93-4	Potential carcinogen	Insoluble Sinker
Ethylene dichloride	Aliphatics, halogenated	6	107-06-2	Flammable w/toxic products Potential carcinogen	Insoluble Sinker
Ethylene oxide	Oxides, alkylene	23	75-21-8	Flammable Corrosive	Soluble Floater
Ethylenediamine	Amines, alkyl	8	107-15-3	Flammable w/toxic products Corrosive	Soluble Floater
Ethylenimine	Amines, alkyl	8	151-56-4	Flammable w/toxic products Poison	Soluble Floater
Formic acid	Acidic compounds, organic	2	64-18-6	Combustible	Soluble Sinker
Furan	Esters and ethers	15	110-00-9	Flammable	Soluble Floater
Furfural	Aldehydes	4	98-01-1	Combustible	Soluble Sinker

(continued)

TABLE 1-1. (continued)

Hazardous Substance	Chemical Class	Guide No.	CAS No.	Hazard(s), in Addition to Toxicity	Behavior in Water
Glycidylaldehyde (Note 2)	Aldehydes	4	765-34-4	Flammable Potential carcinogen	Soluble Sinker
Hexachlorobutadiene	Aliphatics, halogenated	6	87-68-3		Insoluble Sinker
Hexachlorocyclo-pentadiene	Aliphatics, halogenated	6	77-47-4		Insoluble Sinker
Hexachloropropene (Note 2)	Aliphatics, halogenated	6	1888-71-7		Insoluble Sinker
Hexaethyl tetraphosphate	Organophosphates	22	757-58-4	Poison	Soluble Sinker
Hydrazine	Hydrazines and hydrazides	19	302-01-2	Flammable w/toxic products Corrosive Potential carcinogen Poison	Soluble (S.G.=1.00)
Hydrochloric acid	Acidic compounds, inorganic	1	7647-01-0	Corrosive Reactive	Soluble Sinker
Hydrocyanic acid	Cyanides and nitriles	14	74-90-8	Flammable w/toxic products Poison	Soluble Floater
Hydrofluoric acid	Acidic compounds, inorganic	1	7664-39-3	Corrosive Reactive	Soluble Floater
Isobutyl alcohol	Alcohols and glycols	3	78-83-1	Flammable	Soluble Floater
Isophorone	Ketones	20	78-59-1	Combustible	Soluble Floater
Isoprene	Aliphatic hydrocarbons	5	78-79-5	Flammable Polymerizable	Insoluble Floater
Isosafrole (Note 2)	Aromatic hydrocarbons	10	120-58-1	Potential carcinogen	Insoluble Sinker
Lead fluoborate (in aqueous solution)	Heavy metals	18	13814-96-5		Soluble Sinker
Malathion	Organophosphates	22	121-75-5		Insoluble Sinker
Mercury	Heavy metals	18	7439-97-6		Insoluble Sinker

(continued)

TABLE 1-1. (continued)

Hazardous Substance	Chemical Class	Guide No.	CAS No.	Hazard(s), in Addition to Toxicity	Behavior in Water
Methacrylonitrile	Cyanides and nitriles	14	126-98-7	Flammable w/toxic products Polymerizable	Soluble Floater
Methanol	Alcohols and glycols	3	67-56-1	Flammable	Soluble Floater
Methapyriline (Note 1)	Amines, aryl	9	91-80-5		Soluble Sinker
Methyl chlorocarbonate	Esters and ethers	15	79-22-1	Flammable w/toxic products	Soluble Sinker
Methyl chloroform	Aliphatics, halogenated	6	71-55-6	Combustible w/toxic products	Insoluble Sinker
Methyl ethyl ketone	Ketones	20	78-93-3	Flammable	Soluble Floater
Methyl hydrazine	Hydrazines and hydrazides	19	60-34-4	Flammable w/toxic products Poison	Soluble Floater
Methyl iodide	Aliphatics, halogenated	6	74-88-4	Potential carcinogen	Soluble Sinker
Methyl isobutyl ketone	Ketones	20	108-10-1	Flammable	Soluble Floater
Methyl isocyanate	Cyanates and isocyanates	13	624-83-9	Flammable w/toxic products	Soluble Floater
Methyl methacrylate	Esters and ethers	15	80-62-6	Flammable Polymerizable	Soluble Floater
Methylene bromide	Aliphatics, halogenated	6	74-95-3		Soluble Sinker
Methylene chloride	Aliphatics, halogenated	6	75-09-2	Combustible w/toxic products	Soluble Sinker
Mevinphos	Organophosphates	22	7786-34-7	Combustible w/toxic products Poison	Soluble Sinker
Monoethylamine	Amines, alkyl	8	75-04-7	Flammable w/toxic products Corrosive	Soluble Floater

(continued)

TABLE 1-1. (continued)

Hazardous Substance	Chemical Class	Guide No.	CAS No.	Hazard(s), in Addition to Toxicity	Behavior in Water
Nickel carbonyl	Heavy metals	18	13463-39-3	Flammable w/toxic products Reactive Potential carcinogen	Insoluble Sinker
Nicotine and salts	Amines, aryl	9	54-11-5	Combustible w/toxic products Poison	Soluble Sinker
Nitric acid	Acidic compounds, inorganic	1	7697-37-2	Corrosive Reactive Oxidizer	Soluble Sinker
Nitrobenzene	Nitro/nitroso compounds	21	98-95-3	Combustible w/toxic products Poison	Insoluble Sinker
Nitrogen dioxide	Acidic compounds, inorganic	1	10102-44-0	Corrosive Oxidizer Poison	Decomposes (Sinker)
2-Nitropropane	Nitro/nitroso compounds	21	79-46-9	Flammable w/toxic products	Soluble Floater
N-Nitrosodiethanolamine	Nitro/nitroso compounds	21	1116-54-7	Combustible w/toxic products Potential carcinogen	Soluble (S.G.=1.00)
N-Nitrosodiethylamine	Nitro/nitroso compounds	21	55-18-5	Combustible w/toxic products Potential carcinogen	Soluble Floater
N-Nitrosodimethylamine	Nitro/nitroso compounds	21	62-75-9	Flammable w/toxic products Potential carcinogen	Soluble (S.G.=1.00)
N-Nitrosodi-n-propyl-amine	Nitro/nitroso compounds	21	621-64-7	Combustible w/toxic products Potential carcinogen	Insoluble Floater
N-Nitroso-N-methylurea (Note 1)	Nitro/nitroso compounds	21	684-93-5	Combustible w/toxic products Potential carcinogen	Soluble Sinker
N-Nitrosomethylvinyl-amine	Nitro/nitroso compounds	21	4549-40-0	Flammable w/toxic products Potential carcinogen	Soluble Sinker

(continued)

TABLE 1-1. (continued)

Hazardous Substance	Chemical Class	Guide No.	CAS No.	Hazard(s), in Addition to Toxicity	Behavior in Water
N-Nitrosopyrrolidine	Nitro/nitroso compounds	21	930-55-2	Combustible w/toxic products Potential carcinogen	Soluble Sinker
Octamethylpyrophosphor-amide	Amides, anilides, and imides	7	152-16-9		Soluble Sinker
Paraldehyde	Esters and ethers	15	123-63-7	Flammable	Soluble Floater
Parathion	Organophosphates	22	56-38-2	Poison	Insoluble Sinker
Pentachloroethane	Aliphatics, halogenated	6	76-01-7	Combustible w/toxic products	Insoluble Sinker
1,3-Pentadiene	Aliphatic hydrocarbons	5	504-60-9	Flammable	Insoluble Floater
Phosphoric acid	Acidic compounds, inorganic	1	7664-38-2	Corrosive	Soluble Sinker
Phosphorus oxychloride	Halides, inorganic	17	10025-87-3	Corrosive Reactive	Decomposes (Sinker)
Phosphorus trichloride	Halides, inorganic	17	7719-12-2	Corrosive Reactive	Decomposes (Sinker)
2-Picoline	Amines, aryl	9	109-06-8	Combustible w/toxic products	Soluble Floater
Polychlorinated biphenyls	Aromatics, halogenated	11	1336-36-3	Potential carcinogen	Insoluble Sinker
Propargite	Sulfates and sulfites	26	2312-35-8	Flammable w/toxic products	Insoluble Sinker
Propargyl alcohol	Alcohols and glycols	3	107-19-7	Flammable Poison	Soluble Floater
Propionic acid	Acidic compounds, organic	2	79-09-4	Combustible	Soluble Floater
Propionic anhydride	Acidic compounds, organic	2	123-62-6	Combustible	Decomposes (Sinker)

(continued)

TABLE 1-1. (continued)

Hazardous Substance	Chemical Class	Guide No.	CAS No.	Hazard(s), in Addition to Toxicity	Behavior in Water
n-Propylamine	Amines, alkyl	8	107-10-8	Flammable w/toxic products	Soluble Floater
Propylene dichloride	Aliphatics, halogenated	6	78-87-5	Flammable w/toxic products	Insoluble Sinker
Propylene oxide	Oxides, alkylene	23	75-56-9	Flammable Corrosive	Soluble Floater
1,2-Propylenimine	Amines, alkyl	8	75-55-8	Flammable w/toxic products Potential carcinogen	Soluble Floater
Pyridine	Amines, aryl	8	110-86-1	Flammable w/toxic products	Soluble Floater
Quinoline	Amines, aryl	9	91-22-5	Combustible	Soluble Sinker
Safrole	Aromatic hydrocarbons	10	94-59-7	Combustible Potential carcinogen	Insoluble Sinker
Selenious acid	Acidic compounds, inorganic	1	7783-00-8		Soluble Sinker
Sodium hypochlorite (in aqueous solution)	Basic compounds	12	10022-70-5 7681-52-9		Soluble Sinker
Styrene	Aromatic hydrocarbons	10	100-42-5	Flammable Polymerizable	Insoluble Floater
Sulfur monochloride	Halides, inorganic	17	12771-08-3	Corrosive Reactive	Decomposes (Sinker)
Sulfuric acid	Acidic compounds, inorganic	1	7664-93-9	Corrosive Reactive	Soluble Sinker
2,4,5-T amines	Amines, aryl	9	2008-46-0	Combustible w/toxic products	Insoluble Sinker
2,4,5-T esters	Esters and ethers	15	N.A.		Insoluble Sinker
1,1,1,2-Tetrachloroethane	Aliphatics, halogenated	6	630-20-6		Insoluble Sinker
1,1,2,2-Tetrachloroethane	Aliphatics, halogenated	6	79-34-5		Insoluble Sinker

(continued)

TABLE 1-1. (continued)

Hazardous Substance	Chemical Class	Guide No.	CAS No.	Hazard(s), in Addition to Toxicity	Behavior in Water
Tetrachloroethylene	Aliphatics, halogenated	6	127-18-4		Insoluble Sinker
Tetraethyldithiopyrophosphate	Organophosphates	22	3689-24-5	Poison	Insoluble Sinker
Tetraethyl lead	Heavy metals	18	78-00-2	Combustible w/toxic products Poison	Insoluble Sinker
Tetraethyl pyrophosphate	Organophosphates	22	107-49-3	Combustible w/toxic products Poison	Soluble Sinker
Tetrahydrofuran	Esters and ethers	15	109-99-9	Flammable	Soluble Floater
Tetranitromethane	Nitro/nitroso compounds	21	509-14-8	Explosive Oxidizer	Insoluble Sinker
Thiophenol	Phenols and cresols	25	108-98-5		Insoluble Sinker
Toluene	Aromatic hydrocarbons	10	108-88-3	Flammable	Insoluble Floater
Toluene diisocyanate	Cyanates and isocyanates	13	584-84-9	Combustible w/toxic products Poison	Decomposes (Sinker)
2,4,5-TP acid esters	Esters and ethers	15	N.A.	Combustible w/toxic products	Insoluble Sinker
1,2,4-Trichlorobenzene	Aromatics, halogenated	11	120-82-1	Combustible w/toxic products	Insoluble Sinker
1,1,2-Trichloroethane	Aliphatics, halogenated	6	79-00-5	Combustible w/toxic products	Insoluble Sinker
Trichloroethylene	Aliphatics, halogenated	6	79-01-6	(Flammable w/toxic products)	Insoluble Sinker
Trichloromonofluoromethane	Aliphatics, halogenated	6	75-69-4		Insoluble Sinker
Triethylamine	Amines, alkyl	8	121-44-8	Flammable w/toxic products Corrosive	Soluble Floater

(continued)

TABLE 1-1. (continued)

Hazardous Substance	Chemical Class	Guide No.	CAS No.	Hazard(s), in Addition to Toxicity	Behavior in Water
Vinyl acetate	Esters and ethers	15	108-05-4	Flammable Polymerizable	Soluble Floater
Vinylidene chloride	Aliphatics, halogenated	6	75-35-4	Flammable w/toxic products Polymerizable Potential carcinogen	Insoluble Sinker
Xylene	Aromatic hydrocarbons	10	1330-20-7	Flammable	Insoluble Floater
Xylenol	Phenols and cresols	25	1300-71-6	Combustible	Soluble Sinker

Note 1: Specific gravity calculated according to Grains Method Given in the Handbook of Chemical Property Estimation Methods (Ref. 1).

Note 2: Specific Gravity and/or solubility estimated based on active chemical groups and physical characteristics of structurally similar compounds.

LEGEND:

N.A. = Not Available
S.G. = Specific Gravity

TABLE 1-2. CERCLA Liquids Numerical Index

DOT ID No.	Hazardous Substance	Chemical Class	Guide No.	CAS No.*	Hazard(s), in Addition to Toxicity	Behavior in Water
1036	Monoethylamine	Amines, alkyl	8	75-04-7	Flammable w/toxic products Corrosive	Soluble Floater
1040	Ethylene oxide	Oxides, alkylene	23	75-21-8	Flammable Corrosive	Soluble Floater
1050	Hydrochloric acid (anhydrous)	Acidic compounds, inorganic	1	7647-01-0	Corrosive Reactive	Soluble Sinker
1051	Hydrocyanic acid	Cyanides and nitriles	14	74-90-8	Flammable w/toxic products Poison	Soluble Floater
1052	Hydrofluoric acid (anhydrous)	Acidic compounds inorganic	1	7664-39-3	Corrosive Reactive	Soluble Floater
1067	Nitrogen dioxide	Acidic compounds, inorganic	1	10102-44-0	Corrosive Oxidizer Poison	Decomposes (Sinker)
1089	Acetaldehyde	Aldehydes	4	75-07-0	Flammable Polymerizable	Soluble Floater
1090	Acetone	Ketones	20	67-64-1	Flammable	Soluble Floater
1092	Acrolein	Aldehydes	4	107-02-8	Flammable Polymerizable Poison	Soluble Floater
1093	Acrylonitrile	Cyanides and nitriles	14	107-13-1	Flammable w/toxic products Polymerizable Potential carcinogen Poison	Soluble Floater
1098	Allyl alcohol	Alcohols and glycols	3	107-18-6	Flammable Poison	Soluble Floater
1100	Allyl chloride	Aliphatics, halogenated	6	107-05-1	Flammable w/toxic products	Insoluble Floater
1104	Amyl acetate	Esters and ethers	15	628-63-7	Flammable	Insoluble Floater
1114	Benzene	Aromatic hydrocarbons	10	71-43-2	Flammable Potential carcinogen	Insoluble Floater

*Chemical Abstract Services Number

(continued)

TABLE 1-2. (continued)

DOT ID No.	Hazardous Substance	Chemical Class	Guide No.	CAS No.	Hazard(s), in Addition to Toxicity	Behavior in Water
1120	1-Butanol	Alcohols and glycols	3	71-36-3	Flammable	Soluble Floater
1123	Butyl acetate	Esters and ethers	15	123-86-4	Flammable	Soluble Floater
1125	Butylamine	Amines, alkyl	8	109-73-9	Flammable w/toxic products	Soluble Floater
1131	Carbon disulfide	Sulfides and mercaptans	27	75-15-0	Flammable w/toxic products	Insoluble Sinker
1134	Chlorobenzene	Aromatics, halogenated	11	108-90-7	Flammable w/toxic products	Insoluble Sinker
1143	Crotonaldehyde	Aldehydes	4	4170-30-3 123-73-9	Flammable	Soluble Floater
1145	Cyclohexane	Aliphatic hydrocarbons	5	110-82-7	Flammable	Insoluble Floater
1150	1,2-trans-Dichloroethylene	Aliphatics, halogenated	6	156-60-5	Flammable w/toxic products	Insoluble Sinker
1154	Diethylamine	Amines,alkyl	8	109-89-7	Flammable w/toxic products Corrosive	Soluble Floater
1155	Ethyl ether	Esters and ethers	15	60-29-7	Explosive, upon standing Flammable	Soluble Floater
1163	1,1-Dimethylhydrazine	Hydrazines and hydrazides	19	57-14-7	Flammable w/toxic products Corrosive Potential carcinogen Poison	Soluble Floater
1165	1,4-Dioxane	Esters and ethers	15	123-91-1	Flammable Potential carcinogen	Soluble Sinker
1173	Ethyl acetate	Esters and ethers	15	141-78-6	Flammable	Soluble Floater
1175	Ethylbenzene	Aromatic hydrocarbons	10	100-41-4	Flammable	Insoluble Floater
1184	Ethylene dichloride	Aliphatics, halogenated	6	107-06-2	Flammable w/toxic products Potential carcinogen	Insoluble Sinker

(continued)

TABLE 1-2. (continued)

DOT ID No.	Hazardous Substance	Chemical Class	Guide No.	CAS No.	Hazard(s), in Addition to Toxicity	Behavior in Water
1185	Ethylenimine	Amines, alkyl	8	151-56-4	Flammable w/toxic products Poison	Soluble Floater
1193	Methyl ethyl ketone	Ketones	20	78-93-3	Flammable	Soluble Floater
1199	Furfural	Aldehydes	4	98-01-1	Combustible	Soluble Sinker
1212	Isobutyl alcohol	Alcohols and glycols	3	78-83-1	Flammable	Soluble Floater
1218	Isoprene	Aliphatic hydrocarbons	5	78-79-5	Flammable Polymerizable	Insoluble Floater
1230	Methanol	Alcohols and glycols	3	67-56-1	Flammable	Soluble Floater
1238	Methyl chlorocarbonate	Esters and ethers	15	79-22-1	Flammable w/toxic products	Soluble Sinker
1239	Chloromethyl methyl ether	Ethers, halogenated	16	107-30-2	Flammable w/toxic products Potential carcinogen Poison	Insoluble Sinker
1244	Methyl hydrazine	Hydrazines and hydrazides	19	60-34-4	Flammable w/toxic products Poison	Soluble Floater
1245	Methyl isobutyl ketone	Ketones	20	108-10-1	Flammable	Soluble Floater
1247	Methyl methacrylate	Esters and ethers	15	80-62-6	Flammable Polymerizable	Soluble Floater
1259	Nickel carbonyl	Heavy metals	18	13463-39-3	Flammable w/toxic products Reactive Potential carcinogen	Insoluble Sinker
1264	Paraldehyde	Esters and ethers	15	123-63-7	Flammable	Soluble Floater
1277	n-Propylamine	Amines, alkyl	8	107-10-8	Flammable w/toxic products	Soluble Floater
1279	Dichloropropane (all isomers)	Aliphatics, halogenated	6	26638-19-7	Flammable w/toxic products	Insoluble Sinker
1279	Propylene dichloride	Aliphatics, halogenated	6	78-87-5	Flammable w/toxic products	Insoluble Sinker

(continued)

TABLE 1-2. (continued)

DOT ID No.	Hazardous Substance	Chemical Class	Guide No.	CAS No.	Hazard(s), in Addition to Toxicity	Behavior in Water
1280	Propylene oxide	Oxides, alkylene	23	75-56-9	Flammable Corrosive	Soluble Floater
1282	Pyridine	Amines, aryl	9	110-86-1	Flammable w/toxic products	Soluble Floater
1294	Toluene	Aromatic hydrocarbons	10	108-88-3	Flammable	Insoluble Floater
1296	Triethylamine	Amines, alkyl	8	121-44-8	Flammable w/toxic products Corrosive	Soluble Floater
1301	Vinyl acetate	Esters and ethers	15	108-05-4	Flammable Polymerizable	Soluble Floater
1303	Vinylidene chloride	Aliphatics, halogenated	6	75-35-4	Flammable w/toxic products Polymerizable Potential carcinogen	Insoluble Sinker
1307	Xylene	Aromatic hydrocarbons	10	1330-20-7	Flammable	Insoluble Floater
1510	Tetranitromethane	Nitro/nitroso compounds	21	509-14-8	Explosive Oxidizer	Insoluble Sinker
1541	Acetone cyanohydrin	Cyanides and nitriles	14	75-86-5	Combustible w/toxic products Poison	Soluble Floater
1547	Aniline	Amines, aryl	9	62-53-3	Combustible w/toxic products Poison	Soluble Sinker
1560	Arsenic trichloride	Halides, inorganic	17	7784-34-1	Corrosive Reactive Poison	Decomposes (Sinker)
1569	Bromoacetone (Note 2)	Ketones	20	598-31-2	Poison	Soluble Sinker
1589	Cyanogen chloride	Cyanides and nitriles	14	506-77-4	Poison	Soluble Sinker
1591	o-Dichlorobenzene	Aromatics, halogenated	11	95-50-1	Combustible w/toxic products	Insoluble Sinker
1593	Methylene chloride	Aliphatics, halogenated	6	75-09-2	Combustible w/toxic products	Soluble Sinker

(continued)

TABLE 1-2. (continued)

DOT ID No.	Hazardous Substance	Chemical Class	Guide No.	CAS No.	Hazard(s), in Addition to Toxicity	Behavior in Water
1595	Dimethyl sulfate	Sulfates and sulfites	26	77-78-1	Combustible w/toxic products Corrosive Potential carcinogen	Soluble Sinker
1604	Ethylenediamine	Amines, alkyl	8	107-15-3	Flammable w/toxic products Corrosive	Soluble Floater
1605	Ethylene dibromide	Aliphatics, halogenated	6	106-93-4	Potential carcinogen	Insoluble Sinker
1611	Hexaethyl tetraphosphate	Organophosphates	22	757-58-4	Poison	Soluble Sinker
1613	Hydrocyanic acid (in aqueous solution)	Cyanides and nitriles	14	74-90-8	Flammable w/toxic products Poison	Soluble Floater
1648	Acetonitrile	Cyanides and nitriles	14	75-05-8	Flammable w/toxic products	Soluble Floater
1649	Tetraethyl lead	Heavy metals	18	78-00-2	Combustible w/toxic products Poison	Insoluble Sinker
1654 1655 1656 1657 1658 1659	Nicotine and salts	Amines, aryl	9	54-11-5	Combustible w/toxic products Poison	Soluble Sinker
1662	Nitrobenzene	Nitro/nitroso compounds	21	98-95-3	Combustible w/toxic products Poison	Insoluble Sinker
1669	Pentachloroethane	Aliphatics, halogenated	6	76-01-7	Combustible w/toxic products	Insoluble Sinker
1702	1,1,1,2-Tetrachloro-ethane	Aliphatics, halogenated	6	630-20-6		Insoluble Sinker
1702	1,1,2,2-Tetrachloro-ethane	Aliphatics, halogenated	6	79-34-5		Insoluble Sinker
1703 1704	Tetraethyldithiopyro-phosphate	Organophosphates	22	3689-24-5	Poison	Insoluble Sinker

(continued)

TABLE 1-2. (continued)

DOT ID No.	Hazardous Substance	Chemical Class	Guide No.	CAS No.	Hazard(s), in Addition to Toxicity	Behavior in Water
1705	Tetraethyl pyrophosphate	Organophosphates	22	107-49-3	Combustible w/toxic products Poison	Soluble Sinker
1710	Trichloroethylene	Aliphatics, halogenated	6	79-01-6	(Flammable w/toxic products)	Insoluble Sinker
1715	Acetic anhydride	Acidic compounds, organic	2	108-24-7	Combustible Corrosive	Soluble Sinker
1716	Acetyl bromide	Aliphatics, halogenated	6	506-96-7	Flammable w/toxic products Corrosive Reactive	Decomposes (Sinker)
1717	Acetyl chloride	Aliphatics, halogenated	6	75-36-5	Flammable w/toxic products Corrosive Reactive	Decomposes (Sinker)
1730	Antimony pentachloride	Halides, inorganic	17	7647-18-9	Corrosive Reactive	Decomposes (Sinker)
1736	Benzoyl chloride	Aromatics, halogenated	11	98-88-4	Combustible w/toxic products Corrosive	Decomposes (Sinker)
1738	Benzyl chloride	Aromatics, halogenated	11	100-44-7	Combustible w/toxic products Corrosive Reactive	Insoluble Sinker
1754	Chlorosulfonic acid	Acidic compounds, organic	2	7790-94-5	Corrosive Reactive	Soluble Sinker
1760	2,2-Dichloropropionic acid	Acidic compounds, organic	2	75-99-0	Corrosive	Soluble Sinker
1760	Nitric acid	Acidic compounds, inorganic	1	7697-37-2	Corrosive Reactive Oxidizer	Soluble Sinker
1779	Formic acid	Acidic compounds, organic	2	64-18-6	Combustible	Soluble Sinker

(continued)

TABLE 1-2. (continued)

DOT ID No.	Hazardous Substance	Chemical Class	Guide No.	CAS No.	Hazard(s), in Addition to Toxicity	Behavior in Water
1789	Hydrochloric acid (in aqueous solution)	Acidic compounds, inorganic	1	7647-01-0	Corrosive Reactive	Soluble Sinker
1790	Hydrofluoric acid (in aqueous solution)	Acidic compounds, inorganic	1	7664-39-3	Corrosive Reactive	Soluble Floater
1791	Sodium hypochlorite (in aqueous solution)	Basic compounds	12	10022-70-5 7681-52-9		Soluble Sinker
1805	Phosphoric acid	Acidic compounds, inorganic	1	7664-38-2	Corrosive	Soluble Sinker
1809	Phosphorus trichloride	Halides, inorganic	17	7719-12-2	Corrosive Reactive	Decomposes (Sinker)
1810	Phosphorus oxychloride	Halides, inorganic	17	10025-87-3	Corrosive Reactive	Decomposes (Sinker)
1828	Sulfur monochloride	Halides, inorganic	17	12771-08-3	Corrosive Reactive	Decomposes (Sinker)
1830 1831 1832	Sulfuric acid	Acidic compounds, inorganic	1	7664-93-9	Corrosive Reactive	Soluble Sinker
1846	Carbon tetrachloride	Aliphatics, halogenated	6	56-23-5	Potential carcinogen	Insoluble Sinker
1848	Propionic acid	Acidic compounds, organic	2	79-09-4	Combustible	Soluble Floater
1886	Benzal chloride	Aromatics, halogenated	11	98-87-3		Insoluble Sinker
1888	Chloroform	Aliphatics, halogenated	6	67-66-3	Potential carcinogen	Insoluble Sinker
1897	Tetrachloroethylene	Aliphatics, halogenated	6	127-18-4		Insoluble Sinker
1915	Cyclohexanone	Ketones	20	108-94-1	Combustible	Soluble Floater
1916	Bis(2-chloroethyl) ether	Ethers, halogenated	16	111-44-4	Combustible w/toxic products Poison	Soluble Sinker

(continued)

TABLE 1-2. (continued)

DOT ID No.	Hazardous Substance	Chemical Class	Guide No.	CAS No.	Hazard(s), in Addition to Toxicity	Behavior in Water
1917	Ethyl acrylate	Esters and ethers	15	140-88-5	Flammable Polymerizable	Soluble Floater
1918	Cumene	Aromatic hydrocarbons	10	98-82-8	Combustible	Insoluble Floater
1921	1,2-Propylenimine	Amines, alkyl	8	75-55-8	Flammable w/toxic products Potential carcinogen	Soluble Floater
1967	Parathion	Organophosphates	22	56-38-2	Poison	Insoluble Sinker
1986	Propargyl alcohol	Alcohols and glycols	3	107-19-7	Flammable Poison	Soluble Floater
1993	Creosote	Phenols and cresols	25	8001-58-9	Combustible Potential carcinogen	Insoluble Sinker
2021	2-Chlorophenol	Phenols and cresols	25	95-57-8	Combustible w/toxic products	Soluble Sinker
2023	Epichlorohydrin	Oxides, alkylene	23	106-89-8	Flammable w/toxic products	Insoluble Sinker
2029 2030	Hydrazine	Hydrazines and hydrazides	19	302-01-2	Flammable w/toxic products Corrosive Potential carcinogen Poison	Soluble (S.G.=1.00)
2031 2032	Nitric acid	Acidic compounds, inorganic	1	7697-37-2	Corrosive Reactive Oxidizer	Soluble Sinker
2047	1,3-Dichloropropene	Aliphatics, halogenated	6	542-75-6	Flammable w/toxic products	Insoluble Sinker
2047	Dichloropropene (all isomers)	Aliphatics, halogenated	6	26952-23-8	Flammable w/toxic products	Insoluble Sinker
2047	Dichloropropene-dichlo-opropane mixture	Aliphatics, halogenated	6	8003-19-8	Flammable w/toxic products	Insoluble Sinker
2055	Styrene	Aromatic hydrocarbons	10	100-42-5	Flammable Polymerizable	Insoluble Floater
2056	Tetrahydrofuran	Esters and ethers	15	109-99-9	Flammable	Soluble Floater

(continued)

TABLE 1-2. (continued)

DOT ID No.	Hazardous Substance	Chemical Class	Guide No.	CAS No.	Hazard(s), in Addition to Toxicity	Behavior in Water
2075	Chloral	Aldehydes	4	75-87-6	Combustible w/toxic products Corrosive Potential carcinogen	Soluble Sinker
2076	Cresol	Phenols and cresols	25	1319-77-3	Combustible	Soluble Sinker
2078	Toluene diisocyanate	Cyanates and isocyanates	13	584-84-9	Combustible w/toxic products Poison	Decomposes (Sinker)
2116	alpha,alpha-Dimethyl-benzylhydroperoxide	Peroxides	24	80-15-9	Explosive Combustible	Insoluble Sinker
2218	Acrylic acid	Acidic compounds, organic	2	79-10-7	Combustible Corrosive Polymerizable	Soluble Sinker
2224	Benzonitrile	Cyanides and nitriles	14	100-47-0	Combustible w/toxic products	Soluble Sinker
2225	Benzenesulfonyl chloride	Acidic compounds, organic	2	98-09-9	Combustible w/toxic products	Insoluble Sinker
2226	Benzotrichloride	Aromatics, halogenated	11	98-07-7	Combustible w/toxic products Corrosive	Decomposes (Sinker)
2232	Chloroacetaldehyde	Aldehydes	4	107-20-0	Combustible w/toxic products Polymerizable	Soluble Sinker
2249	Bis(chloromethyl) ether	Ethers, halogenated	16	542-88-1	Combustible w/toxic products Potential carcinogen	Soluble Sinker
2261	Xylenol	Phenols and cresols	25	1300-71-6	Combustible	Soluble Sinker
2262	Dimethyl carbamoyl chloride (Note 2)	Acidic compounds, organic	2	79-44-7	Combustible w/toxic products Reactive Potential carcinogen	Soluble Sinker
2277	Ethyl methacrylate	Esters and ethers	15	97-63-2	Flammable Polymerizable	Insoluble Floater

(continued)

TABLE 1-2. (continued)

DOT ID No.	Hazardous Substance	Chemical Class	Guide No.	CAS No.	Hazard(s), in Addition to Toxicity	Behavior in Water
2279	Hexachlorobutadiene	Aliphatics, halogenated	6	87-68-3		Insoluble Sinker
2291	Lead fluoborate (in aqueous solution)	Heavy metals	18	13814-96-5		Soluble Sinker
2313	2-Picoline	Amines, aryl	9	109-06-8	Combustible w/toxic products	Soluble Floater
2315	Polychlorinated biphenyls	Aromatics, halogenated	11	1336-36-3	Potential carcinogen	Insoluble Sinker
2321	1,2,4-Trichlorobenzene	Aromatics, halogenated	11	120-82-1	Combustible w/toxic products	Insoluble Sinker
2337	Thiophenol	Phenols and cresols	25	108-98-5		Insoluble Sinker
2362	1,1-Dichloroethane	Aliphatics, halogenated	6	75-34-3	Flammable w/toxic products	Insoluble Sinker
2382	1,2-Dimethylhydrazine	Hydrazines and hydrazides	19	540-73-8	Flammable w/toxic products Corrosive Potential carcinogen	Soluble Floater
2383	Dipropylamine	Amines, alkyl	8	142-84-7	Flammable w/toxic products	Soluble Floater
2389	Furan	Esters and ethers	15	110-00-9	Flammable	Soluble Floater
2404	Ethyl cyanide	Cyanides and nitriles	14	107-12-0	Flammable w/toxic products	Soluble Floater
2480	Methyl isocyanate	Cyanates and isocyanates	13	624-83-9	Flammable w/toxic products	Soluble Floater
2490	Bis(2-chloroisopropyl)-ether	Ethers, halogenated	16	108-60-1	Combustible w/toxic products	Insoluble Sinker
2496	Propionic anhydride	Acidic compounds, organic	2	123-62-6	Combustible	Decomposes (Sinker)
2515	Bromoform	Aliphatics, halogenated	6	75-25-2		Insoluble Sinker

(continued)

TABLE 1-2. (continued)

DOT ID No.	Hazardous Substance	Chemical Class	Guide No.	CAS No.	Hazard(s), in Addition to Toxicity	Behavior in Water
2550	2-Butanone peroxide (Note 2)	Peroxides	24	1338-23-4	Explosive Oxidizer Combustible	Soluble Sinker
2584	Dodecylbenzenesulfonic acid	Acidic compounds, organic	2	27176-87-0	Combustible w/toxic products	Soluble Sinker
2608	2-Nitropropane	Nitro/nitroso compounds	21	79-46-9	Flammable w/toxic products	Soluble Floater
2622	Glycidylaldehyde (Note 2)	Aldehydes	4	765-34-4	Flammable Potential carcinogen	Soluble Sinker
2644	Methyl iodide	Aliphatics, halogenated	6	74-88-4	Potential carcinogen	Soluble Sinker
2646	Hexachlorocyclo-pentadiene	Aliphatics, halogenated	6	77-47-4		Insoluble Sinker
2656	Quinoline	Amines, aryl	9	91-22-5	Combustible	Soluble Sinker
2664	Methylene bromide	Aliphatics, halogenated	6	74-95-3		Soluble Sinker
2672	Ammonium hydroxide	Basic compounds	12	1336-21-6	Corrosive	Soluble Floater
2683	Ammonium sulfide (in aqueous solution)	Sulfides and mercaptans	27	12135-76-1	Flammable w/toxic products	Soluble Sinker
2765	2,4,5-TP acid esters	Esters and ethers	15	N.A.	Combustible w/toxic products	Insoluble Sinker
2765	Propargite	Sulfates and sulfites	26	2312-35-8	Flammable w/toxic products	Insoluble Sinker
2765	2,4,5-T esters	Esters and ethers	15	N.A.		Insoluble Sinker
2765	2,4,5-T amines	Amines, aryl	9	2008-46-0	Combustible w/toxic products	Insoluble Sinker
2783	Dichlorvos	Organophosphates	22	62-73-7	Poison	Insoluble Sinker
2783	Malathion	Organophosphates	22	121-75-5		Insoluble Sinker
2783	Diazinon	Organophosphates	22	333-41-5	Combustible w/toxic products	Insoluble Sinker

(continued)

TABLE 1-2. (continued)

DOT ID No.	Hazardous Substance	Chemical Class	Guide No.	CAS No.	Hazard(s), in Addition to Toxicity	Behavior in Water
2784	Tetraethyl pyrophosphate	Organophosphates	22	107-49-3	Combustible w/toxic products Poison	Soluble Sinker
2783 2784	Parathion	Organophosphates	22	56-38-2	Poison	Insoluble Sinker
2783	Ethion	Organophosphates	22	563-12-2	Poison	Insoluble Sinker
2783	Mevinphos	Organophosphates	22	7786-34-7	Combustible w/toxic products Poison	Soluble Sinker
2789	Acetic acid	Acidic compounds, organic	2	64-19-7	Combustible Corrosive	Soluble Sinker
2809	Mercury	Heavy metals	18	7439-97-6		Insoluble Sinker
2820	Butyric acid	Acidic compounds, organic	2	107-92-6	Combustible	Soluble Floater
2831	1,1,2-Trichloroethane	Aliphatics, halogenated	6	79-00-5	Combustible w/toxic products	Insoluble Sinker
2831	Methyl chloroform	Aliphatics, halogenated	6	71-55-6	Combustible w/toxic products	Insoluble Sinker
2872	1,2-Dibromo-3-chloropropane (Note 2)	Aliphatics, halogenated	6	96-12-8	Combustible w/toxic products Potential carcinogen	Insoluble Sinker
2924	1,4-Dichloro-2-butene	Aliphatics, halogenated	6	764-41-0	Flammable w/toxic products	Insoluble Sinker
9095	Di-n-butylphthalate	Esters and ethers	15	84-74-2		Insoluble Sinker

Note 1: Specific gravity calculated according to Grains Method given in the Handbook of Chemical Property Estimation Methods (Ref. 1).

Note 2: Specific gravity and/or solubility estimated based on active chemical groups and physical characteristics of structurally similar compounds.

LEGEND:
N.A. = Not Available
S.G. = Specific Gravity

SORBENT SELECTION AND USE GUIDE

GUIDE NUMBER 1

CHEMICAL CLASS: Acidic Compounds, Inorganic **REPRESENTATIVE HAZARDOUS LIQUID:** Nitric Acid

For releases onto land, recommended sorbents are listed in prioritized order under the Landspill Scenario, and, for landfill applications, under Landfill Scenario. For additional sorbent prioritization information, see the Sorbent Prioritization tables and accompanying text in Section 9 of the Manual. Also, for additional sorbent specifics turn to the Sorbent Data Sheet found on the page listed in the prioritized sorbent column. Pertinent logistical information, such as application and collection methodology, is given in Section 3 of the Manual.

LANDSPILL - SMALL

Prioritized Sorbents/Type/Pg	Application	Collection	Limitations
(1)Foamed Glass*/pl/2-15	Throw	Pitchfork	R,P,DGC,RT
(2)Ex Mineral/p/2-14	Shovel	Shovel	R,I,W,P,DGC
(2)Foamed Glass*/p/2-15	Shovel	Shovel	R,W,P,DGC
( )			
( )			
( )			

LANDSPILL - MEDIUM

Prioritized Sorbents/Type/Pg	Application	Collection	Limitations
(1)Ex Mineral/p/2-14	Blower	Skiploader	R,I,W,P,DGC
(2)Foamed Glass*/p/2-15	Blower	Skiploader	R,W,P,DGC
(3)Foamed Glass*/pl/2-15	Throw	Skiploader	R,P,DGC,RT
( )			
( )			
( )			

LANDFILL

Prioritized Sorbents/Type/Pg	Estimated** Disposal Costs($)
(1)Ex Mineral/p/2-14	1168-1501
(2)Foamed Glass*/p/2-15	3375
( )	
( )	
( )	
( )	

---

* Not compatible with hydrofluoric acid.
** Cost includes sorbent, drums, mixing, transportation, and disposal associated with landfill of 1000 gals. of hazardous liquid.

Legend:

b	= Boom	pl	= Pillow
CLP	= Cross linked polymer	R	= Not reusable
DGC	= Not effective where ground cover is dense	RT	= Not effective where terrain is rugged
Ex Mineral	= Expanded mineral	SS	= Not for use within environmentally sensitive sites
I	= Not incinerable	TC	= Treated clay/treated natural organic
m	= Mat	TW	= Treated wood fiber
p	= Particulate	W	= Effectiveness reduced when windy
P	= Effectiveness reduced when rainy		

SORBENT SELECTION AND USE GUIDE

GUIDE NUMBER 2

CHEMICAL CLASS: Acidic Compounds, Organic+ **REPRESENTATIVE HAZARDOUS LIQUID:** Acrylic Acid

For releases onto land, recommended sorbents are listed in prioritized order under the Landspill Scenario, and, for landfill applications, under Landfill Scenario. For additional sorbent prioritization information, see the Sorbent Prioritization tables and accompanying text in Section 9 of the Manual. Also, for additional sorbent specifics turn to the Sorbent Data Sheet found on the page listed in the prioritized sorbent column. Pertinent logistical information, such as application and collection methodology, is given in Section 3 of the Manual.

LANDSPILL - SMALL

Prioritized Sorbents/Type/Pg	Application	Collection	Limitations
(1)Wood Fiber/pl/2-12	Throw	Pitchfork	R,P,DGC,RT
(1)CLP/p/2-19	Shovel	Shovel	R,W,SS
(1)CLP/pl/2-19	Throw	Pitchfork	R,DGC,RT
(2)Sorbent Clay/p/2-10	Shovel	Shovel	R,I,P
(2)Foamed Glass/pl/2-15	Throw	Pitchfork	R,P,DGC,RT
(3)Wood Fiber/p/2-12	Shovel	Shovel	R,W,P,DGC

LANDSPILL - MEDIUM

Prioritized Sorbents/Type/Pg	Application	Collection	Limitations
(1)CLP/p/2-19	Blower	Skiploader	R,W,SS
(2)Polypropylene/p/2-18	Blower	Skiploader	W,SS,DGC
(2)Sorbent Clay/p/2-10	Blower	Skiploader	R,I,P
(3)CLP/pl/2-19	Throw	Skiploader	R,DGC,RT
(3)Polypropylene/m/2-18	Throw	Skiploader	W,SS,DGC
(3)Ex Mineral/p/2-14	Blower	Skiploader	R,I,W,P,DGC

LANDFILL

Prioritized Sorbents/Type/Pg	Estimated* Disposal Costs($)
(1)Ex Mineral/p/2-14	1192-1553
(2)Sorbent Clay/p/2-10	1498-11752
(3)Wood Fiber/p/2-12	1939-2441
(4)Polypropylene/p/2-18	3026-9722
(5)Foamed Glass/p/2-15	3277
(6)TW fiber/p/2-13	9471

---

* Cost includes sorbent, drums, mixing, transportation, and disposal associated with landfill of 1000 gals. of hazardous liquid.

Legend:

b	= Boom
CLP	= Cross linked polymer
DGC	= Not effective where ground cover is dense
Ex Mineral	= Expanded mineral
I	= Not incinerable
m	= Mat
p	= Particulate
P	= Effectiveness reduced when rainy
pl	= Pillow
R	= Not reusable
RT	= Not effective where terrain is rugged
SS	= Not for use within environmentally sensitive sites
TC	= Treated clay/treated natural organic
TW	= Treated wood fiber
W	= Effectiveness reduced when windy

SORBENT SELECTION AND USE GUIDE

GUIDE NUMBER 3

CHEMICAL CLASS: Alcohols & Glycols **REPRESENTATIVE HAZARDOUS LIQUID:** Allyl Alcohol

For releases onto land, recommended sorbents are listed in prioritized order under the Landspill Scenario, and, for landfill applications, under Landfill Scenario. For additional sorbent prioritization information, see the Sorbent Prioritization tables and accompanying text in Section 9 of the Manual. Also, for additional sorbent specifics turn to the Sorbent Data Sheet found on the page listed in the prioritized sorbent column. Pertinent logistical information, such as application and collection methodology, is given in Section 3 of the Manual.

LANDSPILL - SMALL

Prioritized Sorbents/Type/Pg	Application	Collection	Limitations
(1)CLP/p/2-19	Shovel	Shovel	R,W,SS
(1)CLP/pl/2-19	Throw	Shovel	R,DGC,RT
(2)Sorbent Clay/p/2-10	Shovel	Shovel	R,I,P
(3)TW Fiber/pl/2-13	Throw	Pitchfork	DGC,RT
(3)Wood Fiber/pl/2-12	Throw	Pitchfork	R,P,DGC,RT
(4)Foamed Glass/pl/2-15	Throw	Pitchfork	R,P,DGC,RT

LANDSPILL - MEDIUM

Prioritized Sorbents/Type/Pg	Application	Collection	Limitations
(1)CLP/p/2-19	Blower	Skiploader	R,W,SS
(2)Polypropylene/p/2-18	Blower	Skiploader	W,SS,DGC
(2)Sorbent Clay/p/2-10	Blower	Skiploader	R,I,P
(3)Ex Mineral/p/2-14	Blower	Skiploader	R,I,W,P,DGC
(3)Polypropylene/m/2-18	Throw	Skiploader	DGC,RT
(3)Polyurethane/m/2-16	Throw	Skiploader	DGC,RT

LANDFILL

Prioritized Sorbents/Type/Pg	Estimated* Disposal Costs($)
(1)Ex Mineral/p/2-14	1159-1482
(2)Sorbent Clay/p/2-10	1479-11416
(3)Wood Fiber/p/2-12	2007-2542
(4)Polypropylene/p/2-18	2703-8389
(5)Foamed Glass/p/2-15	3109
(6)TW Fiber/p/2-13	8414

---

* Cost includes sorbent, drums, mixing, transportation, and disposal associated with landfill of 1000 gals. of hazardous liquid.

Legend:

b = Boom
CLP = Cross linked polymer
DGC = Not effective where ground cover is dense
Ex Mineral = Expanded mineral
I = Not incinerable
m = Mat
p = Particulate
P = Effectiveness reduced when rainy
pl = Pillow
R = Not reusable
RT = Not effective where terrain is rugged
SS = Not for use within environmentally sensitive sites
TC = Treated clay/treated natural organic
TW = Treated wood fiber
W = Effectiveness reduced when windy

GUIDE NUMBER 4

CHEMICAL CLASS: Aldehydes **REPRESENTATIVE HAZARDOUS LIQUID:** Furfural

For releases onto land, recommended sorbents are listed in prioritized order under the Landspill Scenario, and, for landfill applications, under Landfill Scenario. For additional sorbent prioritization information, see the Sorbent Prioritization tables and accompanying text in Section 9 of the Manual. Also, for additional sorbent specifics turn to the Sorbent Data Sheet found on the page listed in the prioritized sorbent column. Pertinent logistical information, such as application and collection methodology, is given in Section 3 of the Manual.

LANDSPILL - SMALL

Prioritized Sorbents/Type/Pg	Application	Collection	Limitations
(1)CLP/p/2-19	Shovel	Shovel	R,W,SS
(1)CLP/pl/2-19	Throw	Pitchfork	R,DGC,RT
(2)Sorbent Clay/p/2-10	Shovel	Shovel	R,I,P
(2)Foamed Glass/pl/2-15	Shovel	Shovel	R,W,P,DGC
(3)Wood Fiber/pl/2-12	Throw	Pitchfork	R,P,DGC,RT
(3)TW Fiber/pl/2-13	Throw	Pitchfork	DGC,RT

LANDSPILL - MEDIUM

Prioritized Sorbents/Type/Pg	Application	Collection	Limitations
(1)CLP/p/2-19	Blower	Skiploader	R,W,SS
(2)Sorbent Clay/p/2-10	Blower	Skiploader	R,I,P
(2)Polypropylene/p/2-18	Blower	Skiploader	R,SS,DGC
(3)Ex Mineral/p/2-14	Blower	Skiploader	R,I,W,P,DGC
(3)Wood Fiber/p/2-12	Blower	Skiploader	R,W,P,DGC
(3)Polypropylene/m/2-18	Throw	Skiploader	DGC,RT

NON-FLOATING SPILL

LANDFILL

Prioritized Sorbents/Type/Pg	Estimated* Disposal Costs($)
(1)Ex Mineral/p/2-14	1172-1511
(2)Sorbent Clay/p/2-10	1586-13300
(3)Wood Fiber/p/2-12	1851-2311
(4)Polypropylene/p/2-18	2732-8509
(5)Foamed Glass/p/2-15	3430
(6)Diatomite/p/2-11	9331

---

* Cost includes sorbent, drums, mixing, transportation, and disposal associated with landfill of 1000 gals. of hazardous liquid.

Legend:

b = Boom
CLP = Cross linked polymer
DGC = Not effective where ground cover is dense
Ex Mineral = Expanded mineral
I = Not incinerable
m = Mat
p = Particulate
P = Effectiveness reduced when rainy
pl = Pillow
R = Not reusable
RT = Not effective where terrain is rugged
SS = Not for use within environmentally sensitive sites
TC = Treated clay/treated natural organic
TW = Treated wood fiber
W = Effectiveness reduced when windy

SORBENT SELECTION AND USE GUIDE

GUIDE NUMBER 5

CHEMICAL CLASS: Aliphatic Hydrocarbons **REPRESENTATIVE HAZARDOUS LIQUID:** Isoprene

For releases onto land, recommended sorbents are listed in prioritized order under the Landspill Scenario, for water, under the Floating Spill Scenario, and, for landfill applications, under Landfill Scenario. For additional sorbent prioritization information, see the Sorbent Prioritization tables and accompanying text in Section 9 of the Manual. Also, for additional sorbent specifics turn to the Sorbent Data Sheet found on the page listed in the prioritized sorbent column. Pertinent logistical information, such as application and collection methodology, is given in Section 3 of the Manual.

LANDSPILL - SMALL

Prioritized Sorbents/Type/Pg	Application	Collection	Limitations
(1)CLP/p/2-19	Shovel	Shovel	R,W,SS
(1)CLP/pl/2-19	Throw	Pitchfork	R,DGC,RT
(2)Wood Fiber/pl/2-12	Throw	Pitchfork	R,P,DGC,RT
(2)TW Fiber/pl/2-13	Throw	Pitchfork	DGC,RT
(3)Sorbent Clay/p/2-10	Shovel	Shovel	R,I,P
(3)Foamed Glass/pl/2-15	Throw	Pitchfork	R,P,DGC,RT

LANDSPILL - MEDIUM

Prioritized Sorbents/Type/Pg	Application	Collection	Limitations
(1)CLP/p/2-19	Blower	Skiploader	R,W,SS
(2)CLP/pl/2-19	Throw	Skiploader	R,DGC,RT
(3)Sorbent Clay/p/2-10	Blower	Skiploader	R,I,P
(3)Polypropylene/p/2-18	Blower	Skiploader	W,SS,DGC
(4)Ex Mineral/p/2-14	Blower	Skiploader	R,I,W,P,DGC
(4)Polypropylene/m/2-18	Throw	Skiploader	DGC,RT

FLOATING SPILL

NON-FLOWING - SMALL POND

Prioritized Sorbents/Type/Pg	Application	Collection	Limitations
(1)CLP/pl/2-19	Throw	Spear	R
(2)Polyurethane/m/2-16	Throw	Spear	
(3)TW Fiber/pl/2-13	Throw	Spear	
(4)Polypropylene/m/2-18	Throw	Spear	
(5)Polyethylene/m/2-17	Throw	Spear	
( )			

NON-FLOWING - LARGE LAKE

Prioritized Sorbents/Type/Pg	Application	Collection	Limitations
(1)CLP/pl/2-19	Boat/Throw	Boat/Spear	R
(2)Polyurethane/m/2-16	Boat/Throw	Boat/Spear	
(3)TW Fiber/pl/2-13	Boat/Throw	Boat/Spear	
(4)Polypropylene/m/2-18	Boat/Throw	Boat/Spear	
(5)Polyethylene/m/2-17	Boat/Throw	Boat/Spear	
( )			

FLOWING - SMALL CREEK

Prioritized Sorbents/Type/Pg	Application	Collection	Limitations
(1)Polypropylene/b/2-18	Hand	Hand	
(2)Polyurethane/b/2-16	Hand	Hand	
(3)TW Fiber/b/2-13	Hand	Hand	
( )			

FLOWING - LARGE RIVER

Prioritized Sorbents/Type/Pg	Application	Collection	Limitations
(1)Polypropylene/b/2-18	Boat/Hand	Boat/Winch	
(2)Polyurethane/b/2-16	Boat/Hand	Boat/Winch	
(3)TW Fiber/b/2-13	Boat/Hand	Boat/Winch	
( )			

LANDFILL

Prioritized Sorbents/Type/Pg	Estimated* Disposal Costs($)
(1)Ex Mineral/p/2-14	1144-1451
(2)Sorbent Clay/p/2-10	1589-13361
(3)Wood Fiber/p/2-12	1971-2488
(4)Polypropylene/p/2-18	2717-8448
(5)Foamed Glass/p/2-15	3113
(6)CLP/p/2-19	4069

---

* Cost includes sorbent, drums, mixing, transportation, and disposal associated with landfill of 1000 gals. of hazardous liquid.

Legend:

b	= Boom	pl	= Pillow
CLP	= Cross linked polymer	R	= Not reusable
DGC	= Not effective where ground cover is dense	RT	= Not effective where terrain is rugged
Ex Mineral	= Expanded mineral	SS	= Not for use within environmentally sensitive sites
I	= Not incinerable	TC	= Treated clay/treated natural organic
m	= Mat	TW	= Treated wood fiber
p	= Particulate	W	= Effectiveness reduced when windy
P	= Effectiveness reduced when rainy		

SORBENT SELECTION AND USE GUIDE

GUIDE NUMBER 6

CHEMICAL CLASS: Aliphatics, Halogenated+ **REPRESENTATIVE HAZARDOUS LIQUID:** Trichloroethylene

For releases onto land, recommended sorbents are listed in prioritized order under the Landspill Scenario, and, for landfill applications, under Landfill Scenario. For additional sorbent prioritization information, see the Sorbent Prioritization tables and accompanying text in Section 9 of the Manual. Also, for additional sorbent specifics turn to the Sorbent Data Sheet found on the page listed in the prioritized sorbent column. Pertinent logistical information, such as application and collection methodology, is given in Section 3 of the Manual.

LANDSPILL - SMALL

Prioritized Sorbents/Type/Pg	Application	Collection	Limitations
(1)CLP/p/2-18	Shovel	Shovel	R,W,SS
(1)CLP/pl/2-18	Throw	Pitchfork	R,DGC,RT
(2)Wood Fiber/pl/2-12	Throw	Pitchfork	R,P,DGC,RT
(2)TW Fiber/p/2-13	Shovel	Shovel	R,W,DGC
(3)Sorbent Clay/p/2-10	Shovel	Shovel	R,I,P
(3)Foamed Glass/pl/2-15	Throw	Pitchfork	R,P,DGC,RT

LANDSPILL - MEDIUM

Prioritized Sorbents/Type/Pg	Application	Collection	Limitations
(1)CLP/p/2-19	Blower	Skiploader	R,W,SS
(2)CLP/pl/2-19	Throw	Skiploader	R,DGC,RT
(3)Sorbent Clay/p/2-10	Blower	Skiploader	R,I,P2
(3)Polypropylene/p/2-18	Blower	Skiploader	W,SS,DGC
(3)Foamed Glass/pl/2-15	Throw	Skiploader	R,P,DGC,RT
(4)Ex Mineral/p//2-14	Blower	Skiploader	R,I,W,P,DGC

LANDFILL

Prioritized Sorbents/Type/Pg	Estimated* Disposal Costs($)
(1)Ex Mineral/p/2-14	177-1520
(2)Sorbent Clay/p/2-10	1546-12604
(3)Wood Fiber/p/2-12	2051-2606
(4)Polypropylene/p/2-18	3149-10277
(5)CLP/p/2-19	3905
(6)Foamed Glass/p/2-15	3930

\+ Chloroethane floats - the following sorbents are recommended for use when chloroethane is spilled into (1) a non-flowing water body: CLP/pl, TW fiber/pl, or polyurethane/m and into (2) a flowing water body: polyurethane/b, polypropylene/b or TW fiber/b. Allyl chloride floats - the following sorbents are recommended for use when allyl chloride is spilled into (1) a non-flowing water body: CLP/pl, TW fiber/pl, or polypropylene/m; and into (2) a flowing water body: polypropylene/b, polyurethane/b, or TW fiber/b.

* Cost includes sorbent, drums, mixing, transportation, and disposal associated with landfill of 1000 gals. of hazardous liquid.

Legend:

- b = Boom
- CLP = Cross linked polymer
- DGC = Not effective where ground cover is dense
- Ex Mineral = Expanded mineral
- I = Not incinerable
- m = Mat
- p = Particulate
- P = Effectiveness reduced when rainy
- pl = Pillow
- R = Not reusable
- RT = Not effective where terrain is rugged
- SS = Not for use within environmentally sensitive sites
- TC = Treated clay/treated natural organic
- TW = Treated wood fiber
- W = Effectiveness reduced when windy

SORBENT SELECTION AND USE GUIDE

GUIDE NUMBER 7

CHEMICAL CLASS: Amides, Anilides and Imides **REPRESENTATIVE HAZARDOUS LIQUID:** Octamethyl Pyrophosphoramide

INSUFFICIENT DATA

SORBENT SELECTION AND USE GUIDE

GUIDE NUMBER 8

CHEMICAL CLASS: Amines, Alkyl+ **REPRESENTATIVE HAZARDOUS LIQUID:** Triethylamine

For releases onto land, recommended sorbents are listed in prioritized order under the Landspill Scenario, and, for landfill applications, under Landfill Scenario. For additional sorbent prioritization information, see the Sorbent Prioritization tables and accompanying text in Section 9 of the Manual. Also, for additional sorbent specifics turn to the Sorbent Data Sheet found on the page listed in the prioritized sorbent column. Pertinent logistical information, such as application and collection methodology, is given in Section 3 of the Manual.

LANDSPILL - SMALL

Prioritized Sorbents/Type/Pg	Application	Collection	Limitations
(1)CLP/p/2-19	Shovel	Shovel	R,W,SS
(1)CLP/pl/2-19	Throw	Pitchfork	R,DGC,RT
(2)Sorbent Clay/p/2-10	Shovel	Shovel	R,I,P
(3)Wood fiber/pl/2-12	Throw	Pitchfork	R,P,DGC,RT
(3)TW Fiber/pl/2-13	Throw	Pitchfork	DGC,RT
(4)Foamed Glass/pl/2-15	Throw	Pitchfork	R,P,DGC,RT

LANDSPILL - MEDIUM

Prioritized Sorbents/Type/Pg	Application	Collection	Limitations
(1)CLP/p/2-19	Blower	Skiploader	R,W,SS
(2)CLP/pl/2-19	Throw	Skiploader	R,DGC,RT
(3)Sorbent Clay/p/2-10	Blower	Skiploader	R,I,P
(3)Polypropylene/p/2-18	Blower	Skiploader	W,SS,DGC
(4)Ex Mineral/p/2-14	Blower	Skiploader	R,I,W,P,DGC
(4)Polypropylene/m/2-18	Throw	Skiploader	DGC,RT

LANDFILL

Prioritized Sorbents/Type/Pg	Estimated* Disposal Costs($)
(1)Ex Mineral/p/2-14	1211-1595
(2)Sorbent Clay/p/2-10	1546-12604
(3)Wood Fiber/p/2-12	2636-3470
(4)Polypropylene/p/2-18	2686-8318
(5)Foamed Glass/p/2-15	3026
(6)CLP/p/2-19	4866

---

* Cost includes sorbent, drums, mixing, transportation, and disposal associated with landfill of 1000 gals. of hazardous liquid.

Legend:

b	= Boom	pl	= Pillow
CLP	= Cross linked polymer	R	= Not reusable
DGC	= Not effective where ground cover is dense	RT	= Not effective where terrain is rugged
Ex Mineral	= Expanded mineral	SS	= Not for use within environmentally sensitive sites
I	= Not incinerable	TC	= Treated clay/treated natural organic
m	= Mat	TW	= Treated wood fiber
p	= Particulate	W	= Effectiveness reduced when windy
P	= Effectiveness reduced when rainy		

SORBENT SELECTION AND USE GUIDE

GUIDE NUMBER 9

CHEMICAL CLASS: Amines, Aryl

REPRESENTATIVE HAZARDOUS LIQUID: Pyridine

For releases onto land, recommended sorbents are listed in prioritized order under the Landspill Scenario, and, for landfill applications, under Landfill Scenario. For additional sorbent prioritization information, see the Sorbent Prioritization tables and accompanying text in Section 9 of the Manual. Also, for additional sorbent specifics turn to the Sorbent Data Sheet found on the page listed in the prioritized sorbent column. Pertinent logistical information, such as application and collection methodology, is given in Section 3 of the Manual. NOTE: Sorbent prioritization for this guide is based on assessed compatibility and estimated sorption capacity data.

LANDSPILL - SMALL

Prioritized Sorbents/Type/Pg	Application	Collection	Limitations
(1)Wood Fiber/pl/2-12	Throw	Pitchfork	R,P,DGC,RT
(2)Sorbent Clay/p/2-10	Shovel	Shovel	R,I,P
(2)Foamed Glass/pl/2-15	Throw	Pitchfork	R,P,DGC,RT
(3)Wood Fiber/p/2-12	Shovel	Shovel	R,W,P,DGC
(3)TW Fiber/pl/2-13	Throw	Pitchfork	DGC,RT
(4)Polypropylene/m/2-18	Throw	Pitchfork	DGC,RT

LANDSPILL - MEDIUM

Prioritized Sorbents/Type/Pg	Application	Collection	Limitations
(1)Sorbent Clay/p/2-10	Blower	Skiploader	R,I,P
(1)Polypropylene/p/2-18	Blower	Skiploader	W,SS,DGC
(2)Diatomite/p/2-11	Blower	Skiploader	R,I,W,P
(2)Ex Mineral/p/2-14	Blower	Skiploader	R,I,W,P,DGC
(2)Wood Fiber/p/2-12	Blower	Skiploader	R,W,P,DGC
(2)Polypropylene/m/2-18	Throw	Skiploader	DGC,RT

LANDFILL

Prioritized Sorbents/Type/Pg	Estimated* Disposal Costs($)
(1)Ex Mineral/p/2-14	1147-1456
(2)Sorbent Clay/p/2-10	1596-13471
(3)Wood Fiber/p/2-12	1823-2271
(4)Polypropylene/p/2-18	2445-7325
(5)Foamed Glass/p/2-15	3132
(6)Diatomite/p/2-11	6949

---

* Cost includes sorbent, drums, mixing, transportation, and disposal associated with landfill of 1000 gals. of hazardous liquid.

Legend:

- b = Boom
- CLP = Cross linked polymer
- DGC = Not effective where ground cover is dense
- Ex Mineral = Expanded mineral
- I = Not incinerable
- m = Mat
- p = Particulate
- P = Effectiveness reduced when rainy
- pl = Pillow
- R = Not reusable
- RT = Not effective where terrain is rugged
- SS = Not for use within environmentally sensitive sites
- TC = Treated clay/treated natural organic
- TW = Treated wood fiber
- W = Effectiveness reduced when windy

SORBENT SELECTION AND USE GUIDE

GUIDE NUMBER 10

CHEMICAL CLASS: Aromatic Hydrocarbons **REPRESENTATIVE HAZARDOUS LIQUID:** Toluene

For releases onto land, recommended sorbents are listed in prioritized order under the Landspill Scenario, for water, under the Floating Spill Scenario, and, for landfill applications, under Landfill Scenario. Also, for additional sorbent prioritization information, see the Sorbent Prioritization tables and accompanying text in Section 9 of the Manual. Also, for additional sorbent specifics turn to the Sorbent Data Sheet found on the page listed in the prioritized sorbent column. Pertinent logistical information, such as application and collection methodology, is given in Section 3 of the Manual.

LANDSPILL - SMALL

Prioritized Sorbents/Type/Pg	Application	Collection	Limitations
(1)Feathers/pl/2-20	Throw	Pitchfork	DGC,RT
(2)CLP/p/2-19	Shovel	Shovel	R,W,SS
(2)CLP/pl/2-19	Throw	Pitchfork	R,DGC,RT
(3)Sorbent Clay/p/2-10	Shovel	Shovel	R,I,P
(3)TC/p/2-21	Shovel	Shovel	R,I
(4)Wood Fiber/pl/2-12	Throw	Pitchfork	R,P,DGC,RT

LANDSPILL - MEDIUM

Prioritized Sorbents/Type/Pg	Application	Collection	Limitations
(1)CLP/p/2-19	Blower	Skiploader	R/W/SS
(2)TC/p/2-21	Blower	Skiploader	R/I
(3)Sorbent Clay/p/2-10	Blower	Skiploader	R,I,P
(3)Polypropylene/p/2-18	Blower	Skiploader	W,SS,DGC
(3)Feathers/pl/2-20	Throw	Skiploader	DCG,RT
(4)Ex Mineral/p/2-14	Blower	Skiploader	R,I,W,P,DGC

FLOATING SPILL

NON-FLOWING - SMALL POND

Prioritized Sorbents/Type/Pg	Application	Collection	Limitations
(1)CLP/pl/2-19	Throw	Spear	R
(2)Feathers/pl/2-20	Throw	Spear	
(2)TW Fiber/pl/2-13	Throw	Spear	
(3)Polyurethane/m/2-16	Throw	Spear	
(4)Polypropylene/m/2-18	Throw	Spear	
(4)Polyethylene/m/2-17	Throw	Spear	

NON-FLOWING - LARGE LAKE

Prioritized Sorbents/Type/Pg	Application	Collection	Limitations
(1)CLP/pl/2-19	Boat/Throw	Boat/Spear	R
(2)Feathers/pl/2-20	Boat/Throw	Boat/Spear	
(3)TW Fiber/pl/2-13	Boat/Throw	Boat/Spear	
(4)Polypropylene/m/2-18	Boat/Throw	Boat/Spear	
(4)Polyurethane/m/2-16	Boat/Throw	Boat/Spear	
(5)Polyethylene/m/2-17	Boat/Throw	Boat/Spear	

FLOWING - SMALL CREEK

Prioritized Sorbents/Type/Pg	Application	Collection	Limitations
(1)Polypropylene/b/2-18	Hand	Hand	
(2)Polyurethane/b/2-16	Hand	Hand	
(3)Feathers/b/2-20	Hand	Hand	
(4)TW Fiber/b/2-13	Hand	Hand	

FLOWING - LARGE RIVER

Prioritized Sorbents/Type/Pg	Application	Collection	Limitations
(1)Polyproplene/b/2-18	Boat/Hand	Boat/Winch	
(2)Polyurethane/b/2-16	Boat/Hand	Boat/Winch	
(3)Feathers/b/2-20	Baot/Hand	Boat/Winch	
(4)TW Fiber/b/2-13	Boat/Hand	Boat/Winch	

LANDFILL

Prioritized Sorbents/Type/Pg	Estimated* Disposal Costs($)
(1)Ex Mineral/p/2-14	1143-1448
(2)TC/p/2-21	1401-2655
(3)Sorbent Clay/p/2-10	1674-14852
(4)Wood Fiber/p/2-12	1927-2423
(5)Polypropylene/p/2-18	2721-8463
(6)Foamed Glass/p/2-15	3271

---

* Cost includes sorbent, drums, mixing, transportation, and disposal associated with landfill of 1000 gals. of hazardous liquid.

Legend:

b	= Boom	pl	= Pillow
CLP	= Cross linked polymer	R	= Not reusable
DGC	= Not effective where ground cover is dense	RT	= Not effective where terrain is rugged
Ex Mineral	= Expanded mineral	SS	= Not for use within environmentally sensitive sites
I	= Not incinerable	TC	= Treated clay/treated natural organic
m	= Mat	TW	= Treated wood fiber
p	= Particulate	W	= Effectiveness reduced when windy
P	= Effectiveness reduced when rainy		

SORBENT SELECTION AND USE GUIDE

GUIDE NUMBER 11

CHEMICAL CLASS: Aromatics, Halogenated **REPRESENTATIVE HAZARDOUS LIQUID:** o-Dichlorobenzene

For releases onto land, recommended sorbents are listed in prioritized order under the Landspill Scenario, and, for landfill applications, under Landfill Scenario. For additional sorbent prioritization information, see the Sorbent Prioritization tables and accompanying text in Section 9 of the Manual. Also, for additional sorbent specifics turn to the Sorbent Data Sheet found on the page listed in the prioritized sorbent column. Pertinent logistical information, such as application and collection methodology, is given in Section 3 of the Manual.

LANDSPILL - SMALL

Prioritized Sorbents/Type/Pg	Application	Collection	Limitations
(1)CLP/p/2-19	Shovel	Shovel	R,W,SS
(1)CLP/pl/2-19	Throw	Pitchfork	R,DGC,RT
(2)Wood Fiber/pl/2-12	Throw	Pitchfork	R,P,DGC,RT
(3)Sorbent Clay/p/2-10	Shovel	Shovel	R,I,P
(3)Foamed Glass/pl/2-15	Throw	Pitchfork	R,P,DGC,RT
(4)TW Fiber/pl/2-13	Throw	Pitchfork	DGC,RT

LANDSPILL - MEDIUM

Prioritized Sorbents/Type/Pg	Application	Collection	Limitations
(1)CLP/p/2-19	Blower	Skiploader	R,W,SS
(2)Sorbent Clay/p/2-10	Blower	Skiploader	R,I,P
(2)Polypropylene/p/2-18	Blower	Skiploader	W,SS,DGC
(3)Ex Mineral/p/2-14	Blower	Skiploader	R,I,W,P,DGC
(3)Wood Fiber/p/2-12	Blower	Skiploader	R,W,P,DGC
(3)Polypropylene/m/2-18	Throw	Skiploader	DGC,RT

LANDFILL

Prioritized Sorbents/Type/Pg	Estimated* Disposal Costs($)
(1)Ex Mineral/p/2-14	1220-1614
(2)Sorbent Clay/p/2-10	1513-12015
(3)Wood Fiber/p/2-12	1923-2417
(4)Polypropylene/p/2-18	2918-9277
(5)Foamed Glass/p/2-15	3665
(6)CLP/p/2-19	4237

* Cost includes sorbent, drums, mixing, transportation, and disposal associated with landfill of 1000 gals. of hazardous liquid.

Legend:

b	= Boom	pl	= Pillow
CLP	= Cross linked polymer	R	= Not reusable
DGC	= Not effective where ground cover is dense	RT	= Not effective where terrain is rugged
Ex Mineral	= Expanded mineral	SS	= Not for use within environmentally sensitive sites
I	= Not incinerable	TC	= Treated clay/treated natural organic
m	= Mat	TW	= Treated wood fiber
p	= Particulate	W	= Effectiveness reduced when windy
P	= Effectiveness reduced when rainy		

SORBENT SELECTION AND USE GUIDE

GUIDE NUMBER 12

CHEMICAL CLASS: Basic Compounds   **REPRESENTATIVE HAZARDOUS LIQUID:** Ammonium Hydroxide

For releases onto land, recommended sorbents are listed in prioritized order under the Landspill Scenario, and, for landfill applications, under Landfill Scenario. For additional sorbent prioritization information, see the Sorbent Prioritization tables and accompanying text in Section 9 of the Manual. Also, for additional sorbent specifics turn to the Sorbent Data Sheet found on the page listed in the prioritized sorbent column. Pertinent logistical information, such as application and collection methodology, is given in Section 3 of the Manual.

LANDSPILL - SMALL

Prioritized Sorbents/Type/Pg	Application	Collection	Limitations
(1)CLP/p/2-19	Shovel	Shovel	R,W,SS
(1)CLP/pl/2-19	Throw	Pitchfork	R,DGC,RT
(2)Sorbent Clay/p/2-10	Shovel	Shovel	R,I,P
(2)Foamed Glass/pl/2-15	Throw	Pitchfork	R,P,DGC,RT
(3)Ex Mineral/p/2-14	Shovel	Shovel	R,I,W,P,DGC
(4)Foamed Glass/p/2-15	Shovel	Shovel	R,W,P,DGC

LANDSPILL - MEDIUM

Prioritized Sorbents/Type/Pg	Application	Collection	Limitations
(1)CLP/p/2-19	Blower	Skiploader	R,W,SS
(2)Sorbent Clay/p/2-10	Blower	Skiploader	R,I,P
(3)Ex Mineral/p/2-14	Blower	Skiploader	R,I,W,P,DGC
(3)CLP/pl/2-19	Throw	Skiploader	R,DGC,RT
(4)Foamed Glass/p/2-15	Blower	Skiploader	R,W,P,DGC
(4)Foamed Glass/pl/2-15	Throw	Skiploader	R,P,DGC,RT

LANDFILL

Prioritized Sorbents/Type/Pg	Estimated* Disposal Costs($)
(1)Ex Mineral/p/2-14	1175-1518
(2)Sorbent Clay/p/2-10	1530-12315
(3)Foamed Glass/p/2-15	2572
(4)CLP/p/2-19	18784
( )	
( )	

---

* Cost includes sorbent, drums, mixing, transportation, and disposal associated with landfill of 1000 gals. of hazardous liquid.

Legend:

b	= Boom	pl	= Pillow
CLP	= Cross linked polymer	R	= Not reusable
DGC	= Not effective where ground cover is dense	RT	= Not effective where terrain is rugged
Ex Mineral	= Expanded mineral	SS	= Not for use within environmentally sensitive sites
I	= Not incinerable	TC	= Treated clay/treated natural organic
m	= Mat	TW	= Treated wood fiber
p	= Particulate	W	= Effectiveness reduced when windy
P	= Effectiveness reduced when rainy		

SORBENT SELECTION AND USE GUIDE

GUIDE NUMBER 13

CHEMICAL CLASS: Cyanates and Isocyanates **REPRESENTATIVE HAZARDOUS LIQUID:** Toluene Diisocyanate

For releases onto land, recommended sorbents are listed in prioritized order under the Landspill Scenario, and, for landfill applications, under Landfill Scenario. For additional sorbent prioritization information, see the Sorbent Prioritization tables and accompanying text in Section 9 of the Manual. Also, for additional sorbent specifics turn to the Sorbent Data Sheet found on the page listed in the prioritized sorbent column. Pertinent logistical information, such as application and collection methodology, is given in Section 3 of the Manual.

LANDSPILL - SMALL

Prioritized Sorbents/Type/Pg	Application	Collection	Limitations
(1)CLP/p/2-19	Shovel	Shovel	R,W,SS
(1)Wood Fiber/p/2-121	Throw	Pitchfork	R,P,DGC,RT
(1)CLP/pl/2-19	Throw	Pitchfork	R,DGC,RT
(2)Sorbent Clay/p/2-10	Shovel	Shovel	R,I,P
(2)Foamed Glass/pl/2-15	Throw	Pitchfork	R,P,DGC,RT
(3)Wood Fiber/p/2-12	Shovel	Shovel	R,W,P,DGC

LANDSPILL - MEDIUM

Prioritized Sorbents/Type/Pg	Application	Collection	Limitations
(1)CLP/p/2-19	Blower	Skiploader	R,W,SS
(1)CLP/pl/2-19	Throw	Skiploader	R,DGC,RT
(2)Polypropylene/p/2-18	Blower	Skiploader	R,SS,DGC
(3)Ex Mineral/p/2-14	Blower	Skiploader	R,I,W,P,DGC
(3)Wood Fiber/p/2-12	Blower	Skiploader	R,W,P,DGC
(3)Polypropylene/m/2-18	Throw	Skiploader	DGC,RT

LANDFILL

Prioritized Sorbents/Type/Pg	Estimated* Disposal Costs($)
(1)Ex Mineral/p/2-14	1199-1569
(2)Wood Fiber/p/2-12	1783-2212
(3)Polypropylene/p/2-18	2882-9128
(4)Foamed Glass/p/2-15	3083
(5)TW Fiber/p/2-13	9954
(6)CLP/p/2-19	18476

---

* Cost includes sorbent, drums, mixing, transportation, and disposal associated with landfill of 1000 gals. of hazardous liquid.

Legend:

b	= Boom	pl	= Pillow
CLP	= Cross linked polymer	R	= Not reusable
DGC	= Not effective where ground cover is dense	RT	= Not effective where terrain is rugged
Ex Mineral	= Expanded mineral	SS	= Not for use within environmentally sensitive sites
I	= Not incinerable	TC	= Treated clay/treated natural organic
m	= Mat	TW	= Treated wood fiber
p	= Particulate	W	= Effectiveness reduced when windy
P	= Effectiveness reduced when rainy		

SORBENT SELECTION AND USE GUIDE

GUIDE NUMBER 14

CHEMICAL CLASS: Cyanides and Nitriles **REPRESENTATIVE HAZARDOUS LIQUID:** Acetonitrile

For releases onto land, recommended sorbents are listed in prioritized order under the Landspill Scenario, for water, under the Floating Spill Scenario, and, for landfill applications, under Landfill Scenario. For additional sorbent prioritization information, see the Sorbent Prioritization tables and accompanying text in Section 9 of the Manual. Also, for additional sorbent specifics turn to the Sorbent Data Sheet found on the page listed in the prioritized sorbent column. Pertinent logistical information, such as application and collection methodology, is given in Section 3 of the Manual.

LANDSPILL - SMALL

Prioritized Sorbents/Type/Pg	Application	Collection	Limitations
(1)CLP/pl/2-19	Throw	Pitchfork	R,DGC,RT
(1)CLP/p/2-19	Shovel	Shovel	R,W,SS
(2)Sorbent Clay/p/2-10	Shovel	Shovel	R,I,P
(2)Foamed Glass/pl/2-15	Throw	Pitchfork	R,P,DGC,RT
(3)Wood Fiber/pl/2-12	Throw	Pitchfork	R,P,DGC,RT
(3)TW Fiber/pl/2-13	Throw	Pitchfork	DGC,RT

LANDSPILL - MEDIUM

Prioritized Sorbents/Type/Pg	Application	Collection	Limitations
(1)Sorbent Clay/p/2-10	Blower	Skiploader	R,I,P
(2)CLP/p/2-19	Blower	Skiploader	R,W,SS
(3)Polypropylene/p/2-18	Blower	Skiploader	R,SS,DGC
(4)Ex Mineral/p/2-14	Blower	Skiploader	R,I,W,P,DGC
(5)Wood Fiber/p/2-12	Blower	Skiploader	R,W,P,DGC
(6)Diatomite/p/2-11	Blower	Skiploader	R,I,W,P

FLOATING SPILL

NON-FLOWING - SMALL POND

Prioritized Sorbents/Type/Pg	Application	Collection	Limitations
(1)CLP/pl/2-19	Throw	Spear	R
(2)TW Fiber/pl/2-13	Throw	Spear	
(3)Polypropylene/m/2-18	Throw	Spear	
(4)Polyethylene/m/2-17	Throw	Spear	
( )			
( )			

NON-FLOWING - LARGE LAKE

Prioritized Sorbents/Type/Pg	Application	Collection	Limitations
(1)CLP/pl/2-19	Boat/Throw	Boat/Spear	R
(2)TW Fiber/pl/2-13	Boat/Throw	Boat/Spear	
(3)Polypropylene/m/2-18	Boat/Throw	Boat/Spear	
(4)Polyethylene/m/2-17	Boat/Throw	Boat/Spear	
( )			
( )			

FLOWING - SMALL CREEK

Prioritized Sorbents/Type/Pg	Application	Collection	Limitations
(1)Polypropylene/b/2-18	Hand	Hand	
(2)TW Fiber/b/2-13	Hand	Hand	
( )			
( )			

FLOWING - LARGE RIVER

Prioritized Sorbents/Type/Pg	Application	Collection	Limitations
(1)Polypropylene/b/2-18	Boat/Hand	Boat/Winch	
(2)TW Fiber/b/2-13	Boat/Hand	Boat/Winch	
( )			
( )			

LANDFILL

Prioritized Sorbents/Type/Pg	Estimated* Disposal Costs($)
(1)Ex Mineral/p/2-14	1163-1491
(2)Sorbent Clay/p/2-10	1521-12158
(3)Wood Fiber/p/2-12	1830-2281
(4)Polypropylene/p/2-18	2493-7525
(5)Foamed Glass/p/2-15	2782
(6)TW Fiber/p/2-13	9694

---

* Cost includes sorbent, drums, mixing, transportation, and disposal associated with landfill of 1000 gals. of hazardous liquid.

Legend:

- b = Boom
- CLP = Cross linked polymer
- DGC = Not effective where ground cover is dense
- Ex Mineral = Expanded mineral
- I = Not incinerable
- m = Mat
- p = Particulate
- P = Effectiveness reduced when rainy
- pl = Pillow
- R = Not reusable
- RT = Not effective where terrain is rugged
- SS = Not for use within environmentally sensitive sites
- TC = Treated clay/treated natural organic
- TW = Treated wood fiber
- W = Effectiveness reduced when windy

GUIDE NUMBER 15

CHEMICAL CLASS: Esters and Ethers **REPRESENTATIVE HAZARDOUS LIQUID:** 1,4-Dioxane

For releases onto land, recommended sorbents are listed in prioritized order under the Landspill Scenario, for water, under the Floating Spill Scenario, and, for landfill applications, under Landfill Scenario. For additional sorbent prioritization information, see the Sorbent Prioritization tables and accompanying text in Section 9 of the Manual. Also, for additional sorbent specifics turn to the Sorbent Data Sheet found on the page listed in the prioritized sorbent column. Pertinent logistical information, such as application and collection methodology, is given in Section 3 of the Manual.

LANDSPILL - SMALL

Prioritized Sorbents/Type/Pg	Application	Collection	Limitations
(1)CLP/p/2-19	Shovel	Shovel	R,W,SS
(1)CLP/pl/2-19	Throw	Pitchfork	R,DGC,RT
(2)Sorbent Clay/p/2-10	Shovel	Shovel	R,I,P
(3)Wood Fiber/p/2-12	Shovel	Shovel	R,W,P,DGC
(3)Wood Fiber/pl/2-12	Throw	Pitchfork	R,P,DGC,RT
(3)TW Fiber/pl/2-13	Throw	Pitchfork	DGC,RT

LANDSPILL - MEDIUM

Prioritized Sorbents/Type/Pg	Application	Collection	Limitations
(1)CLP/p/2-19	Blower	Skiploader	R,W,SS
(2)CLP/pl/2-19	Throw	Skiploader	R,DGC,RT
(3)Sorbent Clay/p/2-10	Blower	Skiploader	R,I,P
(3)Polypropylene/p/2-18	Blower	Skiploader	W,SS,DGC
(4)Ex Mineral/p/2-14	Blower	Skiploader	R,I,W,P,DGC
(4)Wood Fiber/p/2-12	Blower	Skiploader	R,W,P,DGC

FLOATING SPILL

NON-FLOWING - SMALL POND

Prioritized Sorbents/Type/Pg	Application	Collection	Limitations
(1)TW Fiber/pl/2-13	Throw	Spear	
(1)CLP/pl/2-19	Throw	Spear	R
(2)Polyurethane/m/2-16	Throw	Spear	
(3)Polyproplyene/m/2-18	Throw	Spear	
(4)Polyethylene/m/2-17	Throw	Spear	
( )			

NON-FLOWING - LARGE LAKE

Prioritized Sorbents/Type/Pg	Application	Collection	Limitations
(1)CLP/pl/2-19	Boat/Throw	Boat/Spear	R
(2)Polyurethane/m/2-16	Boat/Throw	Boat/Spear	
(3)TW Fiber/pl/2-13	Boat/Throw	Boat/Spear	
(4)Polypropylene/m/2-18	Boat/Throw	Boat/Spear	
(5)Polyethylene/m/2-17	Boat/Throw	Boat/Spear	
( )			

FLOWING - SMALL CREEK

Prioritized Sorbents/Type/Pg	Application	Collection	Limitations
(1)Polypropylene/b/2-18	Hand	Hand	
(2)Polyurethane/b/2-16	Hand	Hand	
(3)TW fiber/b/2-13	Hand	Hand	
( )			

FLOWING - LARGE RIVER

Prioritized Sorbents/Type/Pg	Application	Collection	Limitations
(1)Polypropylene/b/2-19	Boat/Hand	Boat/Winch	
(2)Polyurethane/b/2-17	Boat/Hand	Boat/Winch	
(3)TW Fiber/b/2-13	Boat/Hand	Boat/Winch	
( )			

LANDFILL

Prioritized Sorbents/Type/Pg	Estimated* Disposal Costs($)
(1)Ex Mineral/p/2-14	1140-1440
(2)Sorbent Clay/p/2-10	1564-12911
(3)Wood Fiber/p/2-12	1742-2151
(4)Polypropylene/p/2-15	2784-8723
(5)Foamed Glass/p/2-15	2947
(6)CLP/p/2-19	8379

---

* Cost includes sorbent, drums, mixing, transportation, and disposal associated with landfill of 1000 gals. of hazardous liquid.

Legend:

- b = Boom
- CLP = Cross linked polymer
- DGC = Not effective where ground cover is dense
- Ex Mineral = Expanded mineral
- I = Not incinerable
- m = Mat
- p = Particulate
- P = Effectiveness reduced when rainy
- pl = Pillow
- R = Not reusable
- RT = Not effective where terrain is rugged
- SS = Not for use within environmentally sensitive sites
- TC = Treated clay/treated natural organic
- TW = Treated wood fiber
- W = Effectiveness reduced when windy

SORBENT SELECTION AND USE GUIDE

GUIDE NUMBER 16

CHEMICAL CLASS: Ethers, Halogenated **REPRESENTATIVE HAZARDOUS LIQUID:** 2-Chloroethyl Vinyl Ether

For releases onto land, recommended sorbents are listed in prioritized order under the Landspill Scenario, and, for landfill applications, under Landfill Scenario. For additional sorbent prioritization information, see the Sorbent Prioritization tables and accompanying text in Section 9 of the Manual. Also, for additional sorbent specifics turn to the Sorbent Data Sheet found on the page listed in the prioritized sorbent column. Pertinent logistical information, such as application and collection methodology, is given in Section 3 of the Manual. NOTE: Sorbent prioritization for this guide is based on assessed compatibility and estimated sorption capacity data.

LANDSPILL - SMALL

Prioritized Sorbents/Type/Pg	Application	Collection	Limitations
(1)Sorbent Clay/p/2-10	Shovel	Shovel	R,I,P
(2)Wood Fiber/p/2-12	Shovel	Shovel	R,W,P,DGC
(2)Wood Fiber/pl/2-12	Throw	Pitchfork	R,P,DGC,RT
(2)TW Fiber/pl/2-13	Throw	Pitchfork	DGC,RT
(3)Foamed Glass/pl/2-15	Throw	Pitchfork	R,P,DGC,RT
(4)Ex Mineral/p/2-14	Shovel	Shovel	R,I,W,P,DGC

LANDSPILL - MEDIUM

Prioritized Sorbents/Type/Pg	Application	Collection	Limitations
(1)Sorbent Clay/p/2-10	Blower	Skiploader	R,I,P
(2)Ex Mineral/p/2-14	Blower	Skiploader	R,I,W,P,DGC
(2)Wood Fiber/p/2-12	Blower	Skiploader	R,W,P,DGC
(3)Foamed Glass/p/2-15	Blower	Skiploader	R,P,DGC,RT
(3)Foamed Glass/pl/2-15	Throw	Skiploader	R,P,DGC,RT
(4)Diatomite/p/2-11	Blower	Skiploader	R,I,W,P

LANDFILL

Prioritized Sorbents/Type/Pg	Estimated* Disposal Costs($)
(1)Ex Mineral/p/2-14	1145-1451
(2)Sorbent Clay/p/2-10	1577-13144
(3)Wood Fiber/p/2-12	1759-2176
(4)Foamed Glass/p/2-15	2987
(5)TW Fiber/p/2-13	9992
(6)Diatomite/p/2-11	10169

---

* Cost includes sorbent, drums, mixing, transportation, and disposal associated with landfill of 1000 gals. of hazardous liquid.

Legend:

b	= Boom	pl	= Pillow
CLP	= Cross linked polymer	R	= Not reusable
DGC	= Not effective where ground cover is dense	RT	= Not effective where terrain is rugged
Ex Mineral	= Expanded mineral	SS	= Not for use within environmentally sensitive sites
I	= Not incinerable	TC	= Treated clay/treated natural organic
m	= Mat	TW	= Treated wood fiber
p	= Particulate	W	= Effectiveness reduced when windy
P	= Effectiveness reduced when rainy		

SORBENT SELECTION AND USE GUIDE

GUIDE NUMBER 17

CHEMICAL CLASS: Halides, Inorganic

REPRESENTATIVE HAZARDOUS LIQUID: Phosphorus Trichloride

For releases onto land, recommended sorbents are listed in prioritized order under the Landspill Scenario, and, for landfill applications, under Landfill Scenario. For additional sorbent prioritization information, see the Sorbent Prioritization tables and accompanying text in Section 9 of the Manual. Also, for additional sorbent specifics turn to the Sorbent Data Sheet found on the page listed in the prioritized sorbent column. Pertinent logistical information, such as application and collection methodology, is given in Section 3 of the Manual.

LANDSPILL - SMALL

Prioritized Sorbents/Type/Pg	Application	Collection	Limitations
(1)CLP/p/2-19	Shovel	Shovel	R,W,SS
(1)CLP/pl/2-19	Throw	Pitchfork	R,DGC,RT
(2)Wood Fiber/p/2-12	Shovel	Shovel	R,W,P,DGC
(2)Wood Fiber/pl/2-12	Throw	Pitchfork	R,P,DGC,RT
(2)Polypropylene/m/2-18	Throw	Pitchfork	DGC,RT
(2)Polyurethane/m/2-16	Throw	Pitchfork	DGC,RT

LANDSPILL - MEDIUM

Prioritized Sorbents/Type/Pg	Application	Collection	Limitations
(1)CLP/p/2-19	Blower	Skiploader	R,W,SS
(2)CLP/pl/2-19	Throw	Skiploader	R,DGC,RT
(3)Polypropylene/p/2-18	Blower	Skiploader	W,SS,DGC
(4)Ex Mineral/p/2-14	Blower	Skiploader	R,I,W,P,DGC
(4)Wood Fiber/p/2-12	Blower	Skiploader	R,W,P,DGC
(4)Polypropylene/m/2-18	Throw	Skiploader	DGC,RT

LANDFILL

Prioritized Sorbents/Type/Pg	Estimated* Disposal Costs($)
(1)Ex Mineral/p/2	1144-1449
(2)Wood Fiber/p/2	2139-2736
(3)Polypropylene/p/2-18	3364-11112
(4)Foamed Glass/p/2	3636
(5)CLP/p/2	3765
( )	

---

* Cost includes sorbent, drums, mixing, transportation, and disposal associated with landfill of 1000 gals. of hazardous liquid.

Legend:

b = Boom
CLP = Cross linked polymer
DGC = Not effective where ground cover is dense
Ex Mineral = Expanded mineral
I = Not incinerable
m = Mat
p = Particulate
P = Effectiveness reduced when rainy
pl = Pillow
R = Not reusable
RT = Not effective where terrain is rugged
SS = Not for use within environmentally sensitive sites
TC = Treated clay/treated natural organic
TW = Treated wood fiber
W = Effectiveness reduced when windy

SORBENT SELECTION AND USE GUIDE

GUIDE NUMBER 18

CHEMICAL CLASS: Heavy Metals+ **REPRESENTATIVE HAZARDOUS LIQUID:** Lead Fluoborate (Aqueous Solution)

For releases onto land, recommended sorbents are listed in prioritized order under the Landspill Scenario, and, for landfill applications, under Landfill Scenario. For additional sorbent prioritization information, see the Sorbent Prioritization tables and accompanying text in Section 9 of the Manual. Also, for additional sorbent specifics turn to the Sorbent Data Sheet found on the page listed in the prioritized sorbent column. Pertinent logistical information, such as application and collection methodology, is given in Section 3 of the Manual.

LANDSPILL - SMALL

Prioritized Sorbents/Type/Pg	Application	Collection	Limitations
(1)Wood fiber/pl/2-12	Throw	Pitchfork	R,P,DGC,RT
(1)CLP/pl/2-19	Throw	Pitchfork	R,DGC,RT
(2)CLP/p/2-19	Shovel	Shovel	R,W,SS
(3)Wood Fiber/p/2-12	Shovel	Shovel	R,W,P,DGC
( )			
( )			

LANDSPILL - MEDIUM

Prioritized Sorbents/Type/Pg	Application	Collection	Limitations
(1)CLP/p/2-19	Blower	Skiploader	R,W,SS
(2)CLP/pl/2-19	Throw	Skiploader	R,DGC,RT
(3)Wood Fiber/p/2-12	Blower	Skiploader	R,W,P,DGC
(4)Wood Fiber/pl/2-12	Throw	Skiploader	R,P,DGC,RT
( )			
( )			

LANDFILL

Prioritized Sorbents/Type/Pg	Estimated* Disposal Costs($)
(1)Wood fiber/p/2-12	1592-1930
(2)CLP/p/2-19	11655
( )	
( )	
( )	
( )	

+ No absorbents have been identified that will sorb mercury.
* Cost includes sorbent, drums, mixing, transportation, and disposal associated with landfill of 1000 gals. of hazardous liquid.

Legend:

b	= Boom	pl	= Pillow
CLP	= Cross linked polymer	R	= Not reusable
DGC	= Not effective where ground cover is dense	RT	= Not effective where terrain is rugged
Ex Mineral	= Expanded mineral	SS	= Not for use within environmentally sensitive sites
I	= Not incinerable	TC	= Treated clay/treated natural organic
m	= Mat	TW	= Treated wood fiber
p	= Particulate	W	= Effectiveness reduced when windy
P	= Effectiveness reduced when rainy		

SORBENT SELECTION AND USE GUIDE

GUIDE NUMBER 19

CHEMICAL CLASS: Hydrazines & Hydrazides **REPRESENTATIVE HAZARDOUS LIQUID:** Methyl Hydrazine

For releases onto land, recommended sorbents are listed in prioritized order under the Landspill Scenario, and, for landfill applications, under Landfill Scenario. For additional sorbent prioritization information, see the Sorbent Prioritization tables and accompanying text in Section 9 of the Manual. Also, for additional sorbent specifics turn to the Sorbent Data Sheet found on the page listed in the prioritized sorbent column. Pertinent logistical information, such as application and collection methodology, is given in Section 3 of the Manual.

LANDSPILL - SMALL

Prioritized Sorbents/Type/Pg	Application	Collection	Limitations
(1)CLP/p/2-19	Shovel	Shovel	R,W,SS
(1)CLP/pl/2-19	Throw	Pitchfork	R,DGC,RT
(2)Sorbent Clay/p/2-10	Shovel	Shovel	R,I,P
(3)Polypropylene/p/2-18	Shovel	Shovel	W,SS,DGC
(4)Diatomite/p/2-11	Shovel	Shovel	R,I,W,P
(4)Polypropylene/m/2-18	Throw	Pitchfork	DGC,RT

LANDSPILL - MEDIUM

Prioritized Sorbents/Type/Pg	Application	Collection	Limitations
(1)CLP/p/2-19	Blower	Skiploader	R,W,SS
(2)Polypropylene/p/2-18	Blower	Skiploader	W,SS,DGC
(3)Polypropylene/m/2-18	Throw	Skiploader	DGC,RT
(4)Sorbent Clay/p/2-10	Blower	Skiploader	R,I,P
(4)Polyethylene/m/2-17	Throw	Skiploader	DGC,RT
(5)Diatomite/p/2-11	Blower	Skiploader	R,I,W,P

LANDFILL

Prioritized Sorbents/Type/Pg	Estimated* Disposal Costs($)
(1)Sorbent Clay/p/2-10	1594-13443
(2)Polypropylene/p/2-18	2887-9148
(3)Diatomite/p/2-11	10788
(4)CLP/p/2-19	21902
( )	
( )	

---

* Cost includes sorbent, drums, mixing, transportation, and disposal associated with landfill of 1000 gals. of hazardous liquid.

Legend:

b	= Boom	pl	= Pillow
CLP	= Cross linked polymer	R	= Not reusable
DGC	= Not effective where ground cover is dense	RT	= Not effective where terrain is rugged
Ex Mineral	= Expanded mineral	SS	= Not for use within environmentally sensitive sites
I	= Not incinerable	TC	= Treated clay/treated natural organic
m	= Mat	TW	= Treated wood fiber
p	= Particulate	W	= Effectiveness reduced when windy
P	= Effectiveness reduced when rainy		

SORBENT SELECTION AND USE GUIDE

GUIDE NUMBER 20

CHEMICAL CLASS: Ketones

REPRESENTATIVE HAZARDOUS LIQUID: Methyl Isobutyl Ketone

For releases onto land, recommended sorbents are listed in prioritized order under the Landspill Scenario, for water, under the Floating Spill Scenario, and, for landfill applications, under Landfill Scenario. For additional sorbent prioritization information, see the Sorbent Prioritization tables and accompanying text in Section 9 of the Manual. Also, for additional sorbent specifics turn to the Sorbent Data Sheet found on the page listed in the prioritized sorbent column. Pertinent logistical information, such as application and collection methodology, is given in Section 3 of the Manual.

LANDSPILL - SMALL

Prioritized Sorbents/Type/Pg	Application	Collection	Limitations
(1)CLP/p/2-19	Shovel	Shovel	R,W,SS
(1)CLP/pl/2-19	Throw	Pitchfork	R,DGC,RT
(2)Sorbent Clay/p/2-10	Shovel	Shovel	R,I,P
(3)Wood Fiber/pl/2-12	Throw	Pitchfork	R,P,DGC,RT
(3)TW Fiber/pl/2-13	Throw	Pitchfork	DGC,RT
(4)Foamed Glass/pl/2-15	Throw	Pitchfork	R,P,DGC,RT

LANDSPILL - MEDIUM

Prioritized Sorbents/Type/Pg	Application	Collection	Limitations
(1)CLP/p/2-19	Blower	Skiploader	R,W,SS
(2)CLP/pl/2-19	Throw	Skiploader	R,DGC,RT
(3)Sorbent Clay/p/2-10	Blower	Skiploader	R,I,P
(3)Polypropylene/p/2-18	Blower	Skiploader	R,SS,DGC
(4)Ex Mineral/p/2-14	Blower	Skiploader	R,I,W,P,DGC
(4)Polypropylene/m/2-18	Throw	Skiploader	DGC,RT

FLOATING SPILL

NON-FLOWING - SMALL POND

Prioritized Sorbents/Type/Pg	Application	Collection	Limitations
(1)CLP/pl/2-19	Throw	Spear	
(2)TW Fiber/pl/2-13	Throw	Spear	
(3)Polyurethane/m/2-16	Throw	Spear	
(4)Polyethylene/m/2-17	Throw	Spear	
(5)Polypropylene/m/2-18	Throw	Spear	
( )			

NON-FLOWING - LARGE LAKE

Prioritized Sorbents/Type/Pg	Application	Collection	Limitations
(1)CLP/pl/2-19	Boat/Throw	Boat/Spear	
(2)TW Fiber/pl/2-13	Boat/Throw	Boat/Spear	
(3)Polyurethane/m/2-16	Boat/Throw	Boat/Spear	
(4)Polypropylene/m/2-18	Boat/Throw	Boat/Spear	
(5)Polyethylene/m/2-17	Boat/Throw	Boat/Spear	
( )			

FLOWING - SMALL CREEK

Prioritized Sorbents/Type/Pg	Application	Collection	Limitations
(1)Polypropylene/b/2-18	Hand	Hand	
(2)Polyurethane/b/2-16	Hand	Hand	
(3)TW Fiber/b/2-13	Hand	Hand	
( )			

FLOWING - LARGE RIVER

Prioritized Sorbents/Type/Pg	Application	Collection	Limitations
(1)Polypropylene/b/2-18	Boat/Hand	Boat/Winch	
(2)Polyurethane/b/2-16	Boat/Hand	Boat/Winch	
(3)TW Fiber/b/2-13	Boat/Hand	Boat/Winch	
( )			

LANDFILL

Prioritized Sorbents/Type/Pg	Estimated* Disposal Costs($)
(1)Ex Mineral/p/2-14	1167-1499
(2)Sorbent Clay/p/2-10	1464-11159
(3)Wood Fiber/p/2-10	1980-2502
(4)Polypropylene/p/2-18	3041-9784
(5)Foamed Glass/p/2-12	3899
(6)CLP/p/2-19	5695

---

* Cost includes sorbent, drums, mixing, transportation, and disposal associated with landfill of 1000 gals. of hazardous liquid.

Legend:

b	= Boom	pl	= Pillow
CLP	= Cross linked polymer	R	= Not reusable
DGC	= Not effective where ground cover is dense	RT	= Not effective where terrain is rugged
Ex Mineral	= Expanded mineral	SS	= Not for use within environmentally sensitive sites
I	= Not incinerable	TC	= Treated clay/treated natural organic
m	= Mat	TW	= Treated wood fiber
p	= Particulate	W	= Effectiveness reduced when windy
P	= Effectiveness reduced when rainy		

GUIDE NUMBER 21

CHEMICAL CLASS: Nitro/Nitroso Compounds **REPRESENTATIVE HAZARDOUS LIQUID:** Nitrobenzene

For releases onto land, recommended sorbents are listed in prioritized order under the Landspill Scenario, for water, under the Floating Spill Scenario, and, for landfill applications, under Landfill Scenario. For additional sorbent prioritization information, see the Sorbent Prioritization tables and accompanying text in Section 9 of the Manual. Also, for additional sorbent specifics turn to the Sorbent Data Sheet found on the page listed in the prioritized sorbent column. Pertinent logistical information, such as application and collection methodology, is given in Section 3 of the Manual. NOTE: Sorbent prioritization for this guide is based on assessed compatibility and estimated sorption capacity data.

LANDSPILL - SMALL

Prioritized Sorbents/Type/Pg	Application	Collection	Limitations
(1)Wood Fiber/pl/2-12	Throw	Pitchfork	R,P,DGC,RT
(2)Sorbent Clay/p/2-10	Shovel	Shovel	R,I,P
(2)Foamed Glass/pl/2-15	Throw	Pitchfork	R,P,DGC,RT
(3)Wood Fiber/p/2-12	Shovel	Shovel	R,W,P,DGC,RT
(3)TW Fiber/pl/2-13	Throw	Pitchfork	DGC,RT
(4)Polypropylene/p/2-18	Shovel	Shovel	W,SS,DGC

LANDSPILL - MEDIUM

Prioritized Sorbents/Type/Pg	Application	Collection	Limitations
(1)Sorbent Clay/p/2-10	Blower	Skiploader	R,I,P
(1)Polypropylene/p/2-18	Blower	Skiploader	W,SS,DGC
(2)Ex Mineral/p/2-14	Blower	Skiploader	R,I,W,P,DGC
(2)Wood Fiber/p/2-12	Blower	Skiploader	R,W,P,DGC,RT
(2)Polypropylene/m/2-18	Throw	Skiploader	DGC,RT
(3)Foamed Glass/p/2-15	Blower	Skiploader	R,P,DGC,RT

FLOATING SPILL

NON-FLOWING - SMALL POND

Prioritized Sorbents/Type/Pg	Application	Collection	Limitations
(1)TW Fiber/pl/2-13	Throw	Spear	
(2)Polypropylene/m/2-18	Throw	Spear	
(3)Polyethylene/m/2-17	Throw	Spear	
( )			
( )			

NON-FLOWING - LARGE LAKE

Prioritized Sorbents/Type/Pg	Application	Collection	Limitations
(1)TW Fiber/pl/2-13	Boat/Throw	Boat/Spear	
(2)Polypropylene/m/2-18	Boat/Throw	Boat/Spear	
(3)Polyethylene/m/2-17	Boat/Throw	Boat/Spear	
( )			
( )			

FLOWING - SMALL CREEK

Prioritized Sorbents/Type/Pg	Application	Collection	Limitations
(1)Polypropylene/b/2-18	Hand	Hand	
(2)TW Fiber/b/2-13	Hand	Hand	
( )			
( )			

FLOWING - LARGE RIVER

Prioritized Sorbents/Type/Pg	Application	Collection	Limitations
(1)Polypropylene/b/2-18	Boat/Hand	Boat/Winch	
(2)TW Fiber/b/2-13	Boat/Hand	Boat/Winch	
( )			
( )			

LANDFILL

Prioritized Sorbents/Type/Pg	Estimated* Disposal Costs($)
(1)Ex Mineral/p/2-14	1209-1591
(2)Sorbent Clay/p/2-10	1764-16426
(3)Wood Fiber/p/2-12	2045-2597
(4)Polypropylene/p/2-18	3005-9633
(5)Foamed Glass/p/2-15	3429
(6)TW Fiber/p/2-13	10330

---

* Cost includes sorbent, drums, mixing, transportation, and disposal associated with landfill of 1000 gals. of hazardous liquid.

Legend:

b	= Boom	pl	= Pillow
CLP	= Cross linked polymer	R	= Not reusable
DGC	= Not effective where ground cover is dense	RT	= Not effective where terrain is rugged
Ex Mineral	= Expanded mineral	SS	= Not for use within environmentally sensitive sites
I	= Not incinerable	TC	= Treated clay/treated natural organic
m	= Mat	TW	= Treated wood fiber
p	= Particulate	W	= Effectiveness reduced when windy
P	= Effectiveness reduced when rainy		

SORBENT SELECTION AND USE GUIDE

GUIDE NUMBER 22

CHEMICAL CLASS: Organophosphates **REPRESENTATIVE HAZARDOUS LIQUID:** Malathion

For releases onto land, recommended sorbents are listed in prioritized order under the Landspill Scenario, and, for landfill applications, under Landfill Scenario. For additional sorbent prioritization information, see the Sorbent Prioritization tables and accompanying text in Section 9 of the Manual. Also, for additional sorbent specifics turn to the Sorbent Data Sheet found on the page listed in the prioritized sorbent column. Pertinent logistical information, such as application and collection methodology, is given in Section 3 of the Manual.

LANDSPILL - SMALL

Prioritized Sorbents/Type/Pg	Application	Collection	Limitations
(1)CLP/p/2-19	Shovel	Shovel	R,W,SS
(1)CLP/pl/2-19	Throw	Pitchfork	R,DGC,RT
(1)Wood Fiber/pl/2-12	Throw	Pitchfork	R,P,DGC,RT
(2)Foamed Glass/pl/2-15	Shovel	Shovel	R,W,P,DGC
(2)Sorbent Clay/p/2-10	Shovel	Shovel	R,I,P
(3)Wood Fiber/p/2-12	Shovel	Shovel	R,W,P,DGC

LANDSPILL - MEDIUM

Prioritized Sorbents/Type/Pg	Application	Collection	Limitations
(1)CLP/p/2-19	Blower	Skiploader	R,W,SS
(2)Sorbent Clay/p/2-10	Blower	Skiploader	R,I,P
(2)Polypropylene/p/2-18	Blower	Skiploader	R,SS,DGC
(3)Ex Mineral/p/2-14	Blower	Skiploader	R,I,W,P,DGC
(3)Wood Fiber/p/2-12	Blower	Skiploader	R,W,P,DGC
(3)Polypropylene/m/2-18	Throw	Skiploader	DGC,RT

LANDFILL

Prioritized Sorbents/Type/Pg	Estimated* Disposal Costs($)
(1)Ex Mineral/p/2-14	1260-1701
(2)Sorbent Clay/p/2-10	1571-13036
(3)Wood Fiber/p/2-12	1956-2467
(4)Polypropylene/p/2-18	3307-10878
(5)Foamed Glass/p/2-15	3841
(6)TW Fiber/p/2-13	10922

---

* Cost includes sorbent, drums, mixing, transportation, and disposal associated with landfill of 1000 gals. of hazardous liquid.

Legend:

b = Boom
CLP = Cross linked polymer
DGC = Not effective where ground cover is dense
Ex Mineral = Expanded mineral
I = Not incinerable
m = Mat
p = Particulate
P = Effectiveness reduced when rainy
pl = Pillow
R = Not reusable
RT = Not effective where terrain is rugged
SS = Not for use within environmentally sensitive sites
TC = Treated clay/treated natural organic
TW = Treated wood fiber
W = Effectiveness reduced when windy

SORBENT SELECTION AND USE GUIDE

GUIDE NUMBER 23

CHEMICAL CLASS: Oxides, Alkylene

REPRESENTATIVE HAZARDOUS LIQUID: Propylene Oxide

For releases onto land, recommended sorbents are listed in prioritized order under the Landspill Scenario, and, for landfill applications, under Landfill Scenario. For additional sorbent prioritization information, see the Sorbent Prioritization tables and accompanying text in Section 9 of the Manual. Also, for additional sorbent specifics turn to the Sorbent Data Sheet found on the page listed in the prioritized sorbent column. Pertinent logistical information, such as application and collection methodology, is given in Section 3 of the Manual.

LANDSPILL - SMALL

Prioritized Sorbents/Type/Pg	Application	Collection	Limitations
(1)CLP/p/2-19	Shovel	Shovel	R,W,SS
(1)CLP/pl/2-19	Throw	Pitchfork	R,DGC,RT
(2)Sorbent Clay/p/2-10	Shovel	Shovel	R,I,P
(3)Wood Fiber/pl/2-12	Throw	Pitchfork	R,P,DGC,RT
(3)TW Fiber/pl/2-13	Throw	Pitchfork	DGC,RT
(4)Foamed Glass/pl/2-15	Throw	Pitchfork	R,P,DGC,RT

LANDSPILL - MEDIUM

Prioritized Sorbents/Type/Pg	Application	Collection	Limitations
(1)CLP/p/2-19	Blower	Skiploader	R,W,SS
(2)CLP/pl/2-19	Throw	Skiploader	R,DGC,RT
(3)Sorbent Clay/p/2-10	Blower	Skiploader	R,I,P
(3)Polypropylene/p/2-18	Blower	Skiploader	W,SS,DGC
(4)Ex Mineral/p/2-14	Blower	Skiploader	R,I,W,P,DGC
(4)Polypropylene/m/2-18	Throw	Skiploader	DGC,RT

LANDFILL

Prioritized Sorbents/Type/Pg	Estimated* Disposal Costs($)
(1)Ex mineral/p/2-14	1177-1522
(2)Sorbent Clay/p/2-10	1638-14210
(3)Wood Fiber/p/2-12	2021-2562
(4)Polypropylene/p/2-18	2725-8480
(5)Foamed Glass/p/2-15	3238
(6)CLP/p/2-19	6446

---

* Cost includes sorbent, drums, mixing, transportation, and disposal associated with landfill of 1000 gals. of hazardous liquid.

Legend:

- b = Boom
- CLP = Cross linked polymer
- DGC = Not effective where ground cover is dense
- Ex Mineral = Expanded mineral
- I = Not incinerable
- m = Mat
- p = Particulate
- P = Effectiveness reduced when rainy
- pl = Pillow
- R = Not reusable
- RT = Not effective where terrain is rugged
- SS = Not for use within environmentally sensitive sites
- TC = Treated clay/treated natural organic
- TW = Treated wood fiber
- W = Effectiveness reduced when windy

SORBENT SELECTION AND USE GUIDE

GUIDE NUMBER 24

CHEMICAL CLASS: Peroxides **REPRESENTATIVE HAZARDOUS LIQUID:** alpha,alpha-Dimethylbenzylhydroperoxide

For releases onto land, recommended sorbents are listed in prioritized order under the Landspill Scenario, and, for landfill applications, under Landfill Scenario. For additional sorbent prioritization information, see the Sorbent Prioritization tables and accompanying text in Section 9 of the Manual. Also, for additional sorbent specifics turn to the Sorbent Data Sheet found on the page listed in the prioritized sorbent column. Pertinent logistical information, such as application and collection methodology, is given in Section 3 of the Manual.

LANDSPILL - SMALL

Prioritized Sorbents/Type/Pg	Application	Collection	Limitations
(1)CLP/p/2-19	Shovel	Shovel	R,W,SS
(1)Wood Fiber/pl/2-12	Throw	Pitchfork	R,P,DGC,RT
(1)CLP/pl/2-19	Throw	Pitchfork	R,DGC,RT
(2)Sorbent Clay/p/2-10	Shovel	Shovel	R,I,P
(2)Foamed Glass/pl/2-15	Throw	Pitchfork	R,P,DGC,RT
(3)Wood Fiber/p/2-12	Shovel	Shovel	R,W,P,DGC

LANDSPILL - MEDIUM

Prioritized Sorbents/Type/Pg	Application	Collection	Limitations
(1)CLP/p/2-19	Blower	Skiploader	R,W,SS
(2)Sorbent Clay/p/2-10	Blower	Skiploader	R,I,P
(2)Polypropylene/p/2-18	Blower	Skiploader	W,SS,DGC
(3)Ex Mineral/p/2-14	Blower	Skiploader	R,I,W,P,DGC
(3)Wood Fiber/p/2-12	Blower	Skiploader	R,W,P,DGC
(3)Polypropylene/m/2-18	Throw	Skiploader	DGC,RT

LANDFILL

Prioritized Sorbents/Type/Pg	Estimated* Disposal Costs($)
(1)Ex Mineral/p/2-14	1172-1510
(2)Sorbent Clay/p/2-10	1564-12911
(3)Wood Fiber/p/2-12	1643-2004
(4)Polypropylene/p/2-18	2825-8892
(5)Foamed Glass/p/2-15	2963
(6)Diatomite/p/2-11	9385

---

* Cost includes sorbent, drums, mixing, transportation, and disposal associated with landfill of 1000 gals. of hazardous liquid.

Legend:

b	= Boom	pl	= Pillow
CLP	= Cross linked polymer	R	= Not reusable
DGC	= Not effective where ground cover is dense	RT	= Not effective where terrain is rugged
Ex Mineral	= Expanded mineral	SS	= Not for use within environmentally sensitive sites
I	= Not incinerable	TC	= Treated clay/treated natural organic
m	= Mat	TW	= Treated wood fiber
p	= Particulate	W	= Effectiveness reduced when windy
P	= Effectiveness reduced when rainy		

SORBENT SELECTION AND USE GUIDE

GUIDE NUMBER 25

CHEMICAL CLASS: Phenols and Cresols **REPRESENTATIVE HAZARDOUS LIQUID:** Cresol

For releases onto land, recommended sorbents are listed in prioritized order under the Landspill Scenario, and, for landfill applications, under Landfill Scenario. For additional sorbent prioritization information, see the Sorbent Prioritization tables and accompanying text in Section 9 of the Manual. Also, for additional sorbent specifics turn to the Sorbent Data Sheet found on the page listed in the prioritized sorbent column. Pertinent logistical information, such as application and collection methodology, is given in Section 3 of the Manual.

LANDSPILL - SMALL

Prioritized Sorbents/Type/Pg	Application	Collection	Limitations
(1)CLP/p/2-19	Shovel	Shovel	R,W,SS
(1)CLP/pl/2-19	Throw	Pitchfork	R,DGC,RT
(1)Wood Fiber/pl/2-12	Throw	Pitchfork	R,P,DGC,RT
(2)Foamed Glass/pl/2-15	Shovel	Shovel	R,W,P,DGC
(2)Sorbent Clay/p/2-10	Shovel	Shovel	R,I,P
(3)Wood Fiber/p/2-12	Shovel	Shovel	R,W,P,DGC

LANDSPILL - MEDIUM

Prioritized Sorbents/Type/Pg	Application	Collection	Limitations
(1)CLP/p/2-19	Blower	Skiploader	R,W,SS
(2)CLP/pl/2-19	Throw	Skiploader	R,DGC,RT
(3)Polypropylene/p/2-18	Blower	Skiploader	R,SS,DGC
(3)Sorbent Clay/p/2-10	Blower	Skiploader	R,I,P
(4)Wood fiber/p/2-12	Blower	Skiploader	R,W,P,DGC
(4)Ex Mineral/p/2-14	Blower	Skiploader	R,I,W,P,DGC

LANDFILL

Prioritized Sorbents/Type/Pg	Estimated* Disposal Costs($)
(1)Ex Mineral/p/2-14	1166-1498
(2)Sorbent Clay/p/2-10	1459-11060
(3)Wood Fiber/p/2-12	1643-2004
(4)Polypropylene/p/2-18	2689-8331
(5)Foamed Glass/p/2-15	3050
(6)CLP/p/2-19	6544

---

* Cost includes sorbent, drums, mixing, transportation, and disposal associated with landfill of 1000 gals. of hazardous liquid.

Legend:

b = Boom
CLP = Cross linked polymer
DGC = Not effective where ground cover is dense
Ex Mineral = Expanded mineral
I = Not incinerable
m = Mat
p = Particulate
P = Effectiveness reduced when rainy
pl = Pillow
R = Not reusable
RT = Not effective where terrain is rugged
SS = Not for use within environmentally sensitive sites
TC = Treated clay/treated natural organic
TW = Treated wood fiber
W = Effectiveness reduced when windy

SORBENT SELECTION AND USE GUIDE

GUIDE NUMBER 26

CHEMICAL CLASS: Sulfates and Sulfites

REPRESENTATIVE HAZARDOUS LIQUID: Dimethyl Sulfate

For releases onto land, recommended sorbents are listed in prioritized order under the Landspill Scenario, and, for landfill applications, under Landfill Scenario. For additional sorbent prioritization information, see the Sorbent Prioritization tables and accompanying text in Section 9 of the Manual. Also, for additional sorbent specifics turn to the Sorbent Data Sheet found on the page listed in the prioritized sorbent column. Pertinent logistical information, such as application and collection methodology, is given in Section 3 of the Manual.

LANDSPILL - SMALL

Prioritized Sorbents/Type/Pg	Application	Collection	Limitations
(1)CLP/p/2-19	Shovel	Shovel	R,W,SS
(1)Wood Fiber/pl/2-12	Throw	Pitchfork	R,P,DGC,RT
(1)TW Fiber/pl/2-13	Throw	Pitchfork	DGC,RT
(1)CLP/pl/2-19	Throw	Pitchfork	R,DGC,RT
(2)Sorbent Clay/p/2-10	Shovel	Shovel	R,I,P
(2)Foamed Glass/pl/2-15	Thrown	Pitchfork	R,P,DGC,RT

LANDSPILL - MEDIUM

Prioritized Sorbents/Type/Pg	Application	Collection	Limitations
(1)CLP/p/2-19	Blower	Skiploader	R,W,SS
(2)Sorbent Clay/p/2-10	Blower	Skiploader	R,I,P
(2)Polypropylene/p/2-18	Blower	Skiploader	R,SS,DGC
(3)Ex Mineral/p/2-14	Blower	Skiploader	R,I,W,P,DGC
(3)Wood Fiber/p/2-12	Blower	Skiploader	R,W,P,DGC
(3)Polypropylene/m/2-18	Throw	Skiploader	DGC,RT

LANDFILL

Prioritized Sorbents/Type/Pg	Estimated* Disposal Costs($)
(1)Ex Mineral/p/2-14	1187-1542
(2)Sorbent Clay/p/2-10	1627-14025
(3)Wood Fiber/p/2-12	2099-2677
(4)Polypropylene/p/2-18	2965-9471
(5)Foamed Glass/p/2-15	3447
(6)Diatomite/p/2-11	9116

---

* Cost includes sorbent, drums, mixing, transportation, and disposal associated with landfill of 1000 gals. of hazardous liquid.

Legend:

- b = Boom
- CLP = Cross linked polymer
- DGC = Not effective where ground cover is dense
- Ex Mineral = Expanded mineral
- I = Not incinerable
- m = Mat
- p = Particulate
- P = Effectiveness reduced when rainy
- pl = Pillow
- R = Not reusable
- RT = Not effective where terrain is rugged
- SS = Not for use within environmentally sensitive sites
- TC = Treated clay/treated natural organic
- TW = Treated wood fiber
- W = Effectiveness reduced when windy

SORBENT SELECTION AND USE GUIDE

GUIDE NUMBER 27

CHEMICAL CLASS: Sulfides and Mercaptans

REPRESENTATIVE HAZARDOUS LIQUID: Carbon Disulfide

For releases onto land, recommended sorbents are listed in prioritized order under the Landspill Scenario, and, for landfill applications, under Landfill Scenario. For additional sorbent prioritization information, see the Sorbent Prioritization tables on and accompanying text in Section 9 of the Manual. Also, for additional sorbent specifics turn to the Sorbent Data Sheet found on the page listed in the prioritized sorbent column. Pertinent logistical information, such as application and collection methodology, is given in Section 3 of the Manual.

LANDSPILL - SMALL

Prioritized Sorbents/Type/Pg	Application	Collection	Limitations
(1)CLP/p/2-19	Shovel	Shovel	R,W,SS
(1)CLP/pl/2-19	Throw	Pitchfork	R,DGC,RT
(2)Sorbent Clay/p/2-10	Shovel	Shovel	R,I,P
(3)Wood Fiber/p/2-12	Shovel	Shovel	R,W,P,DGC
(3)Wood Fiber/pl/2-12	Throw	Pitchfork	R,P,DGC,RT
(3)Polypropylene/m/2-18	Throw	Pitchfork	W,SS,DGC

LANDSPILL - MEDIUM

Prioritized Sorbents/Type/Pg	Application	Collection	Limitations
(1)CLP/p/2-19	Blower	Skiploader	R,W,SS
(2)CLP/pl/2-19	Throw	Skiploader	R,DGC,RT
(3)Sorbent Clay/p/2-10	Blower	Skiploader	R,I,P
(3)Polypropylene/p/2-18	Blower	Skiploader	W,SS,DGC
(4)Ex Mineral/p/2-14	Blower	Skiploader	R,I,W,P,DGC
(4)Wood Fiber/p/2-12	Blower	Skiploader	R,W,P,DGC

LANDFILL

Prioritized Sorbents/Type/Pg	Estimated* Disposal Costs($)
(1)Ex Mineral/p/2-14	1192-1553
(2)Sorbent Clay/p/2-10	1588-13333
(3)Wood Fiber/p/2-12	1778-2204
(4)Polypropylene/p/2-18	2739-8540
(5)Foamed Glass/p/2-15	3408
(6)CLP/p/2-19	4069

---

* Cost includes sorbent, drums, mixing, transportation, and disposal associated with landfill of 1000 gals. of hazardous liquid.

Legend:

b = Boom
CLP = Cross linked polymer
DGC = Not effective where ground cover is dense
Ex Mineral = Expanded mineral
I = Not incinerable
m = Mat
p = Particulate
P = Effectiveness reduced when rainy
pl = Pillow
R = Not reusable
RT = Not effective where terrain is rugged
SS = Not for use within environmentally sensitive sites
TC = Treated clay/treated natural organic
TW = Treated wood fiber
W = Effectiveness reduced when windy

2. Sorbent Data

This section of the Manual contains the Sorbent Data Sheets, an index for the Sorbent Data Sheets, sorption capacity data and hazardous liquid/water preference indices. Sorbents are classified into 12 major generic classes:

1.	Sorbent Clay (SC)	7.	Polyurethane (PU)
2.	Diatomite (D)	8.	Polyethylene (PE)
3.	Wood Fiber (WF)	9.	Polypropylene (PP)
4.	Treated Wood Fiber (TWF)	10.	Cross-linked Polymer (CLP)
5.	Expanded Mineral (EM)	11.	Feathers (F)
6.	Foamed Glass (FG)	12.	Treated Clay/Treated Natural Organic (TC)

Sorbents were grouped into the 12 generic classes according to their chemical compositions. For example, all sorbents derived from cellulose were placed in the "Wood Fiber" generic sorbent class, while sorbents derived from cellulose treated to render it hydrophobic were placed in the "Treated Wood Fiber" generic class. Each sorbent class is addressed in a Sorbent Data Sheet; the Sheets appear on pages 2-10 through 2-21. Each data sheet lists the generic sorbent class, the commercial trade-names, the manufacturers' names, addresses, and phone numbers, and data on the types of available sorbents, their cost, bulk density and storage requirements. The sorbent types generally available include: particulate, pillows, pads, mats and booms. (These descriptors are consistently used in the Manual although others, such as powder, granular, sheets, and blankets, have been used by other investigators to describe the same sorbent types.) A pillow sorbent typically contains particulate sorbent within an outer bag. Pad and mat sorbents are generally produced in a flat configuration of unitary sorbent material, with the mat being larger in size than the pad. Sheet and blanket sorbents are even larger forms of pad and mat sorbents.

Other information concerning limitations on the use of a generic sorbent for landspills and floating spills (into both non-flowing and flowing water bodies), as applicable, is presented. The limitations include: (1) not effective where ground cover is dense, (2) not incinerable, (3) effectiveness reduced when rainy, (4) not reusable, (5) not effective where terrain is rugged, (6) not for use within environmentally sensitive sites, and (7) effectiveness reduced when windy. (For additional information, see Section 9.) Hazards relative to gross incompatibilities, health and safety, and environmental concerns are also delineated and the sorbent density (no voids) is given.

Because a large number of individual sorbents are marketed and not all could be tested, the sorbents were grouped into classes from which single sorbents, representative of the others in their classes, were chosen for testing. The criteria used for choosing the sorbent representative of a given generic class included: 1) particle size distribution that qualitatively reflected the average of the other member sorbents, and 2) possession of a general form or structure similar to most other members of its class. To a lesser extent, sorbent selection was influenced by availability of sufficient sorbent for testing. The rationale for grouping sorbents in the same generic class is that they can be expected to behave similarly in terms of compatibility with, and sorption capacity for, a given hazardous liquid insofar as they are chemically similar. The surface free energy of a sorbent material (see Section 8) is related to the chemical nature of the solid and is an important factor, although not the only one (pore size and surface area are also included) in influencing sorption capacity. Thus with respect to a specific hazardous liquid, sorption capacities of sorbents within a generic class are expected to be more similar than those of sorbents of different generic classes.

Sorption capacity data for the representative sorbent/hazardous liquid pairs and water are given in Table 2-1. The sorption capacity values are reported in grams of liquid sorbed per gram of sorbent. To convert sorption capacity in units of grams of liquid per gram of sorbent (or weight per unit weight) to pounds of liquid per cubic foot of sorbent, apply the formula (1):

$$\text{Sorption Capacity}\left(\frac{\text{lb liquid}}{\text{cu.ft. sorbent}}\right) = \text{Sorption Capacity}\left(\frac{\text{g liquid}}{\text{g sorbent}}\right) \times \text{Sorbent Bulk Density}\left(\frac{\text{lb}}{\text{cu.ft.}}\right) \qquad (1)$$

For some applications, sorption capacity in pounds of liquid per cubic foot of sorbent may be useful (see Section 5).

Included in Table 2-1 are all available laboratory, literature, and estimated (as described below) sorption capacity data, appropriately indicated as such in the table. If the sorbent/hazardous liquid pair is either known from accepted chemical principles or from observations in the laboratory to be incompatible, this is indicated by an "I". If the sorbent is not wetted by the hazardous liquid, this is indicated by "NW". An observed sorbent/hazardous liquid incompatibility (from project laboratory tests) was indicated by deterioration, degradation, or disintegration of the sorbent, or by bubbling or gas formation beyond initial outgassing. The laboratory values were determined experimentally in accordance with the test protocol described in Section 5 of the Manual. The literature sources of the data enclosed in square brackets ([ ]) in Table 2-1 are as follows: SC-Toluene: R.E. Temple, W.T. Gooding, P.F.Woerner, and G.F. Bennett, "A New Universal Sorbent for Hazardous Spills" in the Proceedings of the 1978 National Conference on Control of Hazardous Material Spills; FG-Nitric acid, FG-Triethylamine, FG-Toluene, FG-Toluene diisocyanate, FG-Phosphorus trichloride, FG-Carbon disulfide: Diamond Shamrock Technical Bulletin SP-LDS-4F; F-Toluene: Sea Clean product literature; TC-Toluene: Radecca product literature.

The estimated values, for Guide Nos. 9, 16, and 23, were based on

TABLE 2-1. SORPTION CAPACITIES OF REPRESENTATIVE SORBENT/HAZARDOUS LIQUID PAIRS

Chemical Class	Guide No.	Representative Hazardous Liquid	GENERIC SORBENT CLASS											
			SC	D	WF	TWF	EM	FG	PU	PE	PP	CLP	F	TC
Acidic compounds, inorganic	1	Nitric acid	I	I	I	I	7.1	[15.2]	I	I	NW	I	ND	ND
Acidic compounds, organic	2	Acrylic acid	1.3	1.4	2.0	2.0	4.8	11.6	I	13.0	9.0	5.9	ND	ND
Alcohols & glycols	3	Allyl alcohol	1.0	1.2	1.4	1.7	4.0	9.3	12.3	12.0	7.9	3.3	ND	ND
Aldehydes	4	Furfural	1.2	1.7	2.3	1.7	5.4	11.5	18.1	12.8	11.0	4.3	ND	ND
Aliphatic hydrocarbons	5	Isoprene	0.7	1.1	1.2	1.2	3.5	7.7	7.3	12.0	6.5	16.3	ND	ND
Aliphatics, halogenated	6	Trichloroethylene	1.6	2.5	2.4	2.2	6.7	12.1	20.4	25.6	11.3	36.9	ND	ND
Amides, anilides, & imides	7	-	ND	ND	ND	ND	ND	ND	ND	ND	ND	ND	ND	ND
Amines, alkyl	8	Triethylamine	0.8	1.0	0.8	1.3	3.0	[8.6]	8.7	12.2	7.1	14.0	ND	ND
Amines, aryl	9	Pyridine	(1)	(2)	(2)	(2)	(5)	(11)	(18)	(21)	(11)	ND	ND	ND
Aromatic hydrocarbons	10	Toluene	[0.8]	2.1	1.6	1.4	4.5	[9.2]	12.6	17.6	8.3	18.3	[6.1]	[21.6]
Aromatics, halogenated	11	o-Dichlorobenzene	1.5	2.3	2.4	2.0	5.2	11.8	19.2	22.3	11.2	29.6	ND	ND
Basic compounds	12	Ammonium hydroxide	1.0	I	I	I	4.1	13.3	NW	NW	NW	3.8	ND	ND
Cyanates & isocyanates	13	Toluene diisocyanate	I	I	2.6	2.1	5.2	[14]	15.3	21.0	10.7	5.3	ND	ND
Cyanides & nitriles	14	Acetonitrile	0.9	1.1	1.6	1.4	3.8	10.5	NW	10.0	8.6	2.7	ND	ND
Esters & ethers	15	1,4-Dioxane	1.1	1.4	2.3	1.8	5.4	12.6	14.9	14.6	9.5	10.5	ND	ND
Ethers, halogenated	16	2-Chloroethyl vinyl ether	(1.1)	(1.4)	(2.3)	(1.8)	(5.4)	(12.6)	I	I	I	I	ND	ND
Halides, inorganic	17	Phosphorus trichloride	I	I	2.4	I	8.1	[14.4]	18.8	26.3	11.1	41.6	ND	ND
Heavy metals	18	Lead fluoborate	I	I	4.0	NW	I	I	NW	NW	NW	10.5	ND	ND
Hydrazines & hydrazides	19	Methyl hydrazine	0.9	1.1	I	I	I	I	I	8.5	7.7	3.2	ND	ND
Ketones	20	Methyl isobutyl ketone	1.0	1.2	1.4	1.3	3.8	6.7	9.0	10.7	6.5	12.7	ND	ND
Nitro/nitroso compounds	21	Nitrobenzene	(1)	(1)	(2)	(2)	(5)	(12)	(14)	(13)	(10)	ND	ND	ND
Organophosphates	22	Malathion	1.3	1.5	2.2	1.9	4.4	10.5	12.6	16.8	8.9	7.3	ND	ND
Oxides, alkylene	23	Propylene oxide	0.8	1.1	1.4	1.1	3.8	8.9	12.4	12.2	7.9	11.4	ND	ND
Peroxides	24	alpha,alpha-Dimethyl-benzylhydroperoxide	1.1	1.5	2.6	1.7	4.8	12.5	25.1	15.8	9.3	3.9	ND	ND
Phenols & cresols	25	Cresol	1.3	1.5	2.6	2.0	4.9	12.0	I	17.2	10.0	13.9	ND	ND
Sulfates & sulfites	26	Dimethyl sulfate	1.3	2.0	2.1	2.3	5.9	13.1	15.5	14.6	11.2	5.2	ND	ND
Sulfides & mercaptans	27	Carbon disulfide	1.3	2.0	2.7	1.9	5.5	[12.6]	14.4	13.1	11.9	30.2	ND	ND
Water			1.0	1.4	3.3	1.0	5.5	12.9	2.4	1.4	0.1	2.5	ND	ND

* The reported sorption capacity data are given in grams of hazardous liquid per gram of sorbent and were measured experimentally in the laboratory except as noted:

( ) = estimated value
[ ] = literature value

LEGEND:

CLP = Cross-Linked Polymer
D = Diatomite
EM = Expanded Mineral
F = Feathers
FG = Foamed Glass
I = Incompatible
ND = No Data
NW = Non-Wetting
PE = Polyethylene
PP = Polypropylene
PU = Polyurethane
SC = Sorbent Clay
TC = Treated Clay/Treated Natural Organic
TWF = Treated Wood Fiber

extrapolations from the experimentally derived data obtained on the same sorbents with other representative hazardous liquids exhibiting relatively similar physical properties. The salient physical properties of the representative hazardous liquid for Guide No. 9, pyridine, and for Guide No. 16, 2-choroethyl vinyl ether, were determined to be most similar, among the liquids tested, to those of 1,4-dioxane (see Table 8-1). The experimentally measured sorption capacities of 1,4-dioxane with the various sorbents tested were then adjusted for the difference between the specific gravity of 1,4-dioxane and the specific gravities of each of the two liquids, and the resultant estimated sorption capacities were entered into Table 2-1. Likewise for Guide No. 23, the representative hazardous liquid, nitrobenzene, possesses physical properties most similar to those of dimethyl sulfate and again the resultant estimated sorption capacities were entered into Table 2-1. (Insufficient physical property data were available for estimating sorption capacities for Guide No. 7.) Care was used in extrapolating in this manner to avoid those hazardous liquids significantly different in physical properties from the hazardous liquid for which the estimate was prepared. Nevertheless, caution should be excercised by the user in relying totally on judgments based on estimated sorption capacity data. Although the estimated values appear reasonable, they have not been verified experimentally. The Sorbent Selection and Use Guides that rely on assessed compatibility and estimated sorption capacity data are duly noted in the guides.

Hazardous liquid/water preference index data are given in Table 2-2. The hazardous liquid/water preference values reported are the ratio of the sorption capacity of a sorbent with a hazardous liquid to that of the same sorbent with water and, as such, are unitless. The preference index is not measured directly on a hazardous liquid/water system. The preference index provides an indication of a sorbent's performance in a waterspill scenario. The higher the hazardous liquid/water preference index, the greater the hydrophobic quality of the sorbent and the greater the likelihood that it will preferentially sorb free or undissolved hazardous liquid instead of water from an aqueous medium. The reader should also refer to Floating Spills - Hydrophobicity in Section 9 of the Manual for additional information. (Note: Table 2-2 does not contain hazardous liquid/water preference index data on feathers and treated clay/treated natural organic sorbent because sorption capacity data with water for the two generic sorbents were not available for calculating the preference index.)

Table 2-3 is an index of sorbent manufacturers, trade names, and generic sorbent categories and is provided to assist the user in locating Sorbent Data Sheets of interest. The index immediately precedes the Sorbent Data Sheets.

TABLE 2-2. REPRESENTATIVE HAZARDOUS LIQUID/WATER PREFERENCE INDICES

Chemical Class	Guide No.	Representative Hazardous Liquid	GENERIC SORBENT CLASS									
			SC	D	WF	TWF	EM	FG	PU	PE	PP	CLP
Acidic compounds, inorganic	1	Nitric acid	**	**	**	**	1.3	[1.2]	**	**	**	**
Acidic compounds, organic	2	Acrylic acid	1.3	1.0	0.6	2.0	0.9	0.9	**	9.3	90.0	2.4
Alcohols & glycols	3	Allyl alcohol	1.0	0.9	0.4	1.7	0.7	0.7	5.1	8.6	79.0	1.3
Aldehydes	4	Furfural	1.2	1.2	0.7	1.7	1.0	0.9	7.5	9.1	110.0	1.7
Aliphatic hydrocarbons	5	Isoprene	0.7	0.8	0.4	1.2	0.6	0.6	3.0	8.6	65.0	6.5
Aliphatics, halogenated	6	Trichloroethylene	1.6	1.8	0.7	2.2	1.2	0.9	8.5	18.3	113.0	14.8
Amides, anilides, & imides	7	**	**	**	**	**	**	**	**	**	**	**
Amines, alkyl	8	Triethylamine	0.8	0.7	0.2	1.3	0.6	[0.7]	3.6	8.7	71.0	5.6
Amines, aryl	9	Pyridine	(1.0)	(1.4)	(0.6)	(2.0)	(0.9)	(0.9)	(7.5)	(15.0)	(110.0)	**
Aromatic hydrocarbons	10	Toluene	[0.8]	1.5	0.5	1.4	0.8	[0.7]	5.3	(12.6)	83.0	7.3
Aromatics, halogenated	11	o-Dichlorobenzene	1.5	1.6	0.7	2.0	1.0	0.9	8.0	15.9	112.0	11.8
Basic compounds	12	Ammonium hydroxide	1.0	**	**	**	0.8	1.0	**	**	**	1.5
Cyanates & isocyanates	13	Toluene diisocyanate	**	**	0.8	2.1	1.0	[1.1]	6.4	15.0	107.0	2.1
Cyanides & nitriles	14	Acetonitrile	0.9	0.8	0.5	1.4	0.7	0.8	**	7.1	86.0	1.1
Esters & ethers	15	1,4-Dioxane	1.1	1.0	0.7	1.8	1.0	1.0	6.2	10.4	95.0	4.2
Ethers, halogenated	16	2-Chloroethyl vinyl ether	(1.1)	(1.0)	(0.7)	(1.8)	(1.0)	(1.0)	**	**	**	**
Halides, inorganic	17	Phosphorus trichloride	**	**	0.7	**	1.5	[1.1]	7.9	18.8	111.0	16.6
Heavy metals	18	Lead fluoborate	**	**	1.2	**	**	**	**	**	**	4.2
Hydrazines & hydrazides	19	Methyl hydrazine	0.9	0.8	**	**	**	**	**	6.1	77.0	1.3
Ketones	20	Methyl isobutyl ketone	1.0	0.9	0.4	1.3	0.7	0.5	3.8	7.6	65.0	5.1
Nitro/nitroso compounds	21	Nitrobenzene	(1.0)	(0.7)	(0.6)	(2.0)	(1.0)	(1.0)	(5.8)	(9.3)	(100.0)	**
Organophosphates	22	Malathion	1.3	1.1	0.7	1.9	0.8	0.8	5.3	12.0	89.0	2.9
Oxides, alkylene	23	Propylene oxide	0.8	0.8	0.4	1.1	0.7	0.7	5.2	8.7	79.0	4.6
Peroxides	24	alpha,alpha-Dimethyl-benzylhydroperoxide	1.1	1.1	0.8	1.7	0.9	1.0	10.5	11.3	93.0	1.6
Phenols & cresols	25	Cresol	1.3	1.1	0.8	2.0	0.9	0.9	**	12.3	100.0	5.6
Sulfates & sulfites	26	Dimethyl sulfate	1.3	1.4	0.6	2.3	1.1	1.0	6.3	10.4	112.0	2.1
Sulfides & mercaptans	27	Carbon disulfide	1.3	1.4	0.8	1.9	1.0	[1.0]	6.0	9.4	119.0	12.1

* The reported preference index data are unitless and were determined from sorption capacity data measured experimentally in the laboratory except as noted:

( ) = estimated value

[ ] = literature value

** = Preference index data are not reported for those sorbent/hazardous liquid pairs determined to be incompatible or non-wetting or for which no data exist (see Table B1).

LEGEND:

CLP = Cross-Linked Polymer	PP = Polypropylene
D = Diatomite	PU = Polyurethane
EM = Expanded Mineral	SC = Sorbent Clay
FG = Foamed Glass	TWF = Treated Wood Fiber
PE = Polyethylene	WF = Wood Fiber

Table 2-3. Sorbent Data Sheet Index

Manufacturer, Trade Name or Generic Sorbent Category	Generic Sorbent Category	Sorbent Data Sheet Page
3M Corporation	Polypropylene	77
3M Oil Sorbent	Polypropylene	77
Absorbs-It	Sorbent Clay	69
American Vermiculite Corp.	Expanded Mineral	73
Anderson's Cob Division	Wood Fiber	71
Andesite of California, Inc.	Sorbent Clay	69
ARG-Dri	Foamed Glass	74
ARG-Dri Sorbent	Polypropylene	77
Bregoil Sponge	Treated Wood Fiber	72
BREG-O-S, Inc.	Treated Wood Fiber	72
Clark-Cutler-McDermott Co.	Polypropylene	77
Clay	Sorbent Clay	69
Clean Dri	Sorbent Clay	69
Clean Up 1	Sorbent Clay	69
Clean Water, Inc.	Treated Expanded Mineral/ Treated Wood Fiber	72/73
Conwed Corporation	Polyethylene	76
Conwed Corporation	Treated Wood Fiber	72
Conwed D-Sorbent	Polyethylene	76
Conwed Standard Sorbent	Treated Wood Fiber	72
Corn Cobs	Wood Fiber	71
Cross-Linked Polymer	-	78
Diamond Shamrock Corp.	Foamed Glass	74
Diatomite	-	70
EMCO, Inc.	Cross-Linked Polymer	78
Enviropoly	Polypropylene	77
Envirosorb Co.	Polypropylene	77
Excel Mineral Co.	Sorbent Clay	69

Table 2-3. (continued)

Manufacturer, Trade Name or Generic Sorbent Category	Generic Sorbent Category	Sorbent Data Sheet Page
Expanded Mineral	-	73
F-12 Fuel Pads	Polypropylene	77
Feathers	-	79
Filter-Media Co.	Expanded Mineral	73
Foamed Glass	-	74
Friction	Sorbent Clay	69
Geotech Development Corp.	Wood Fiber	71
Glass	Foamed Glass	74
Graboil	Polyurethane	75
HAZORB	Foamed Glass	74
High Absorbent Compound	Wood Fiber	71
Imbiber Beads	Cross-Linked Polymer	78
JV Manufacturing Co., Inc.	Foamed Glass	74
JV Manufacturing Co., Inc.	Polypropylene	77
Kitty Litter	Sorbent Clay	69
Klensorb	Treated Clay/Treated Natural Organic	80
LITE-R-COBS	Wood Fiber	71
Locksorb	Treated Clay/Treated Natural Organic	80
Lowe's Inc.	Sorbent Clay	69
Lowe's Safety Absorbent	Sorbent Clay	69
Micafil	Expanded Mineral	73
Oil Dry Corp.	Diatomite	70
Oil Dry Corp.	Sorbent Clay	69
Palabora Vermiculite	Expanded Mineral	73
Perlite	Expanded Mineral	73

Table 2-3. (continued)

Manufacturer, Trade Name or Generic Sorbent Category	Generic Sorbent Category	Sorbent Data Sheet Page
Perlite King	Expanded Mineral	73
Polyethylene	-	76
Polypropylene	-	77
Polyurethane	-	75
Radecca, Inc.	Treated Clay/Treated Natural Organic	80
R.B.H. Cybernetics Ltd.	Polyurethane	75
Safestep	Sorbent Clay	69
Sawdust	Wood Fiber	71
Sea Clean, Inc.	Feathers	79
Sea Clean Oil Sorbent	Feathers	79
SLIKWIK	Wood Fiber	71
SND-G	Sorbent Clay	69
SND-M	Sorbent Clay	69
SND-O	Diatomite	70
Sorbent Clay	-	69
Sorbent Products Co., Inc.	Polypropylene	77
Sorbent Sciences	Polypropylene	77
Sorbent Sciences	Polyurethane	75
SPC Oil Sorbent	Polypropylene	77
SS	Polyurethane	75
Swimming Pool Filtration Media	Diatomite	70

Table 2-3. (continued)

Manufacturer, Trade Name or Generic Sorbent Category	Generic Sorbent Category	Sorbent Data Sheet Page
Treated Clay/Treated Natural Organic	-	80
Treated Wood Fiber	-	72
Vermiculite	Expanded Mineral	73
Waverly	Sorbent Clay	69
Waverly Mineral Products Co.	Sorbent Clay	69
Wood Fiber	-	71

SORBENT DATA SHEET

GENERIC CLASS: Sorbent Clay

SND-G, SNG-M	Clean Dri, Quick Sorb, Absorbs-It Clean Up 1	Safestep, Friction
o Oil Dry Corporation Chicago, IL 312-321-1515 o Cost: Particulate - $0.05/lb o Bulk Density: 29-36 lb/ft3 o Store: Dry	o Excel-Mineral Co. Goleta, CA 805-683-5321 o Cost: Particulate - $0.12/lb o Bulk Density: 34-38 lb/ft3 o Store: Dry	o Andesite of California, Inc o Cost: Particulate - $0.49/lb o Bulk Density: 37 lb/ft3 o Store: Dry
Lowe's Safety Absorbent	**Waverly**	
o Lowe's, Inc. South Ben, In 219-234-8191 o Cost: Particulate $0.04/lb o Bulk Density: 42 lb/ft3 o Store: Dry	o Waverly Mineral Products Co. Bala Cynwyd, PA 215-668-2808 o Cost: Particulate - $1.50/lb o Bulk Density: 30-40 lb/ft3 o Store: Dry	

NOTE: Sorbent Clays are commercially available as kitty litter

LIMITATIONS

Landspill
Not reusable or incinerable, effectiveness reduced when rainy

Floating Spill - Non-Flowing Mode
Not for use on spills into water

Floating Spill - Flowing Mode
Not for use on spills into water

INCOMPATIBILITIES

- o Acidic compounds, inorganic
- o Halides, inorganic
- o Cyanates and isocyanates
- o Heavy metals

HEALTH/SAFETY CONSIDERATIONS
Wear dust respirator and eye protection when handling particulate and other protective equipment appropriate for the hazardous liquid

ENVIRONMENTAL CONSIDERATIONS
Does not biodegrade, but is a natural component of the environment

SORBENT DENSITY (NO VOIDS)*
137.2 pounds/cubic foot

PERFORMANCE

Sorption capacity data and hazardous liquid/water preference index information are given in Tables 2-1 and 2-2, respectively. If the sorbent/hazardous liquid pair is either known, suspected, or observed to be incompatible, an "I" is given in Table 2-1, or if the sorbent is not wet by the hazardous liquid, an "NW" is used.

*This is the density of the solid material without any voids that is used in selected formulae contained in Section 3 and is not to be confused with sorbent bulk density listed above under specific product information.

SORBENT DATA SHEET

GENERIC CLASS: Diatomite

SND-O o Oil Dry Corporation Chicago, IL 312-321-1515 o Cost: Particulate - \$1.45/lb o Bulk Density: 28 lb/ft3 o Store: Dry		

NOTE: Diatomite is used as a filtration medium for swimming pools

LIMITATIONS

<u>Landspill</u>
Not reusable or incinerable; effectiveness reduced when rainy or windy

<u>Floating Spill - Non-Flowing Mode</u>
Not for use on spills into water

<u>Floating Spill - Flowing Mode</u>
Not for use on spills into water

INCOMPATIBILITIES

- o Acidic compounds, inorganic
- o Basic compounds
- o Cyanates and isocyanates
- o Halides, inorganic
- o Heavy metals

HEALTH/SAFETY CONSIDERATIONS
Wear dust respirator, eye protection, and other protective equipment appropriate for the hazardous liquid

ENVIRONMENTAL CONSIDERATIONS
Does not biodegrade but is a natural component of the environment

SORBENT DENSITY (NO VOIDS)*
156.0 pounds/cubic foot

PERFORMANCE

Sorption capacity data and hazardous liquid/water preference index information are given in Tables 2-1 and 2-2, respectively. If the sorbent/hazardous liquid pair is either known, suspected, or observed to be incompatible, an "I" is given in Table 2-1, or if the sorbent is not wet by the hazardous liquid, an "NW" is used.

*This is the density of the solid material without any voids that is used in selected formulae contained in Section 3 and is not to be confused with sorbent bulk density listed above under specific product information.

SORBENT DATA SHEET

GENERIC CLASS: Wood Fiber

Slikwik/Lite-R-Cobs	High Absorbent Compound/Quick Wick Pillows	
o The Andersons Cob Division Maumee, OH 800-472-3220 (inside OH) 800-537-3370 (outside OH) o Cost: Particulate - \$0.10-\$0.20/lb o Bulk Density: 12 lb/ft3 (Slikwick) 34 lb/ft3 (Lite-R-Cobs) o Store: Dry	o Geotech Development Corp. King of Prussia, PA 215-337-8515 o Cost: Particulate - \$0.21/lb Pillows - \$2/lb o Bulk Density: 12 lb/ft3 o Store: Dry	

NOTE: Sawdust can be obtained at lumber yards and sawmills

LIMITATIONS

Landspill

Not reusable; effectiveness reduced when rainy; not effective where groundcover is dense; effectiveness of particulate reduce when rainy; pillows not effective where terrain is rugged

Floating Spill - Non-Flowing Mode

Not for use on spills into water

Floating Spill - Flowing Mode

Not for use on spills into water

INCOMPATIBILITIES

- o Acidic compounds, inorganic
- o Basic compounds
- o Hydrazines and hydrazides

HEALTH/SAFETY CONSIDERATIONS

Wear dust respirator and eye protection when handling particulate and other protective equipment appropriate for the hazardous liquid

ENVIRONMENTAL CONSIDERATIONS

Biodegrades and is a natural component of the environment

SORBENT DENSITY (NO VOIDS)*

49.9 pounds/cubic foot

PERFORMANCE

Sorption capacity data and hazardous liquid/water preference index information are given in Tables 2-1 and 2-2, respectively. If the sorbent/hazardous liquid pair is either known, suspected, or observed to be incompatible, an "I" is given in Table 2-1, or if the sorbent is not wet by the hazardous liquid, an "NW" is used.

*This is the density of the solid material without any voids that is used in selected formulae contained in Section 3 and is not to be confused with sorbent bulk density listed above under specific product information.

SORBENT DATA SHEET

GENERIC CLASS: Treated Wood Fiber

Conwed Standard Sorbent	Bregoil Sponge	
o Conwed Corporation St. Paul, MN 612-221-1144 o Cost: Mat - \$2.50/lb Pillow - \$10/lb Boom - \$3.75/lb o Bulk Density: 3.8 lb/ft3 o Store: Dry, dark	o BREG-O-S, Inc. New York, NY 212-860-3689 o Cost: Particulate - \$1.75/lb Pillow - \$1.48/lb Boom - \$1.55/lb o Bulk Density: 7 lb/ft3 o Store: Dry, dark	

LIMITATIONS

Landspill

Not effective where groundcover is dense; particulate not reusable and effectiveness reduced when windy; pillows not effective where terrain is rugged

Floating Spill - Non-Flowing Mode

None

Floating Spill - Flowing Mode

None

INCOMPATIBILITIES

- o Acidic compounds, inorganic
- o Halides, inorganic
- o Hydrazines and hydrazides

HEALTH/SAFETY CONSIDERATIONS

Wear protective equipment appropriate for the hazardous liquid

ENVIRONMENTAL CONSIDERATIONS

Biodegradable

SORBENT DENSITY (NO VOIDS)*

49.9 pounds/cubic foot

PERFORMANCE

Sorption capacity data and hazardous liquid/water preference index information are given in Tables 2-1 and 2-2, respectively. If the sorbent/hazardous liquid pair is either known, suspected, or observed to be incompatible, an "I" is given in Table 2-1, or if the sorbent is not wet by the hazardous liquid, an "NW" is used.

*This is the density of the solid material without any voids that is used in selected formulae contained in Section 3 and is not to be confused with sorbent bulk density listed above under specific product information.

SORBENT DATA SHEET

GENERIC CLASS: Expanded Mineral

Micafil/Perlite King	Palabora Vermiculite	
o Filter-Media Co. Houston, TX 713-780-900 o Cost: Particulate - \$0.12-\$0.31/lb o Bulk Density: 5-7 lb/ft3 o Store: Dry	o American Vermiculite Corp. Atlanta, GA 404-321-5309 o Cost: Particulate - \$0.14/lb o Bulk Density: 3.5-10 lb/ft3 o Store: Dry	

NOTE: Expanded Minerals are common soil amendments and are available at nurseries

LIMITATIONS

Landspill
Not reusable or incinerable; effectiveness reduced when windy or rainy; not effective where groundcover is dense

Floating Spill - Non-Flowing Mode
Not for use on spills into water

Floating Spill - Flowing Mode
Not for use on spills into water

INCOMPATIBILITIES

o Hydrazines and hyrazides
o Heavy Metals

HEALTH/SAFETY CONSIDERATIONS
Wear dust respirator, eye protection and other protective equipment appropriate for the hazardous liquid

ENVIRONMENTAL CONSIDERATIONS
Does not biodegrade but is a natural component of the environment

SORBENT DENSITY (NO VOIDS)*
149.7 pounds/cubic foot

PERFORMANCE

Sorption capacity data and hazardous liquid/water preference index information are given in Tables 2-1 and 2-2, respectively. If the sorbent/hazardous liquid pair is either known, suspected, or observed to be incompatible, an "I" is given in Table 2-1, or if the sorbent is not wet by the hazardous liquid, an "NW" is used.

*This is the density of the solid material without any voids that is used in selected formulae contained in Section 3 and is not to be confused with sorbent bulk density listed above under specific product information.

SORBENT DATA SHEET

GENERIC CLASS: Foamed Glass

Hazorb	ARG-Dri	
o Diamond Shamrock Corp. Irving, TX 214-659-7000 o Cost: Particulate - $3/lb Pillows - $3.20/lb o Bulk Density: 2 lb/ft3 o Store: Dry	o JV Manufacturing Co., Inc. DePere, WI 414-336-3717 o Cost: Pillows - $4.40/lb o Bulk Density: 2 lb/ft3 o Store: Dry	

LIMITATIONS

Landspill

Not reusable; effectiveness reduced when rainy; not effective where groundcover is dense; effective of particulate reduced when windy; pillows not effective where terrain is rugged

Floating Spill - Non-Flowing Mode

Not for use on spills into water

Floating Spill - Flowing Mode

Not for use on spills into water

INCOMPATIBILITIES

- o Heavy metals
- o Hydrazines and hydrazides

HEALTH/SAFETY CONSIDERATIONS

Wear dust respirator and eye protection when handling particulate and other protective equipment appropriate for the hazardous liquid

ENVIRONMENTAL CONSIDERATIONS

Does not biodegrade but is a natural component of the environment

SORBENT DENSITY (NO VOIDS)*

143.4 pounds/cubic foot

PERFORMANCE

Sorption capacity data and hazardous liquid/water preference index information are given in Tables 2-1 and 2-2, respectively. If the sorbent/hazardous liquid pair is either known, suspected, or observed to be incompatible, an "I" is given in Table 2-1, or if the sorbent is not wet by the hazardous liquid, an "NW" is used.

*This is the density of the solid material without any voids that is used in selected formulae contained in Section 3 and is not to be confused with sorbent bulk density listed above under specific product information.

SORBENT DATA SHEET

GENERIC CLASS: Polyurethane

Graboil	SS	
o R.B.H. Cybernetics Ltd. Victoria, B.C. Canada 604-478-3122 o Cost: Mats - $15/lb Booms - $15/lb o Bulk Density: 1.4 lb/ft3 o Store: Dry, dark, cool	o Spill Control Co. Chino, CA - 714-591-4704 o Cost: Mats - $6.16/lb Booms - $7.83/lb o Bulk Density: 1.8 lb/ft3 (est.) o Store: Dry, dark, cool	

LIMITATIONS

Landspill

Not effective where terrain is rugged or groundcover is dense

Floating Spill - Non-Flowing Mode

None

Floating Spill - Flowing Mode

None

INCOMPATIBILITIES

- o Acidic compounds, inorganic
- o Hydrazines and hydrazides

HEALTH/SAFETY CONSIDERATIONS

Wear protective equipment appropriate for the hazardous liquid

ENVIRONMENTAL CONSIDERATIONS

Not rapidly biodegraded

SORBENT DENSITY (NO VOIDS)*

71.7 pounds/cubic foot

PERFORMANCE

Sorption capacity data and hazardous liquid/water preference index information are given in Tables 2-1 and 2-2, respectively. If the sorbent/hazardous liquid pair is either known, suspected, or observed to be incompatible, an "I" is given in Table 2-1, or if the sorbent is not wet by the hazardous liquid, an "NW" is used.

*This is the density of the solid material without any voids that is used in selected formulae contained in Section 3 and is not to be confused with sorbent bulk density listed above under specific product information.

SORBENT DATA SHEET

GENERIC CLASS: Polyethylene

Conwed D-Sorbent o Conwed Corporation St. Paul, MN 612-221-114 o Cost: Mat - $4.25/lb o Bulk Density: 1.5 lb/ft3 o Store: Dry, dark, cool		

LIMITATIONS

Landspill

Not effective where terrain is rugged or groundcover is dense

Floating Spill - Non-Flowing Mode

None

Floating Spill - Flowing Mode

Not for use on floating spills into streams or rivers - boom not available

INCOMPATIBILITIES

o Acidic compounds, inorganic

HEALTH/SAFETY CONSIDERATIONS

Wear protective equipment appropriate for the hazardous liquid

ENVIRONMENTAL CONSIDERATIONS

Not rapidly biodegraded

SORBENT DENSITY (NO VOIDS)*

58.5 pounds/cubic foot

PERFORMANCE

Sorption capacity data and hazardous liquid/water preference index information are given in Tables 2-1 and 2-2, respectively. If the sorbent/hazardous liquid pair is either known, suspected, or observed to be incompatible, an "I" is given in Table 2-1, or if the sorbent is not wet by the hazardous liquid, an "NW" is used.

*This is the density of the solid material without any voids that is used in selected formulae contained in Section 3 and is not to be confused with sorbent bulk density listed above under specific product information.

SORBENT DATA SHEET

GENERIC CLASS: Polypropylene

3M Oil Sorbent	ARG-Dri Hydrocarbon Sorbent	Enviropoly
o 3M Corporation St. Paul, MN 612-733-1110 o Cost: Particulate - $3.05/lb Mat - $3.30/lb Pillow - $3.40/lb Boom - $4.30/lb o Bulk Density: 3 lb/ft3 o Store: Dry, dark, cool	o JV Manufacturing Co., Inc. DePere, WI 414-336-3717 o Cost: Particulate - $2/lb Mat - $2.95/lb Pillow - $2.17/lb Boom - $2.57/lb o Bulk Density: 3 lb/ft3 o Store: Dry, dark, cool	o Envirosorb Co. Edmonds, WA 206-778-7485 o Cost Mat - $2/lb Boom - $3.95/lb o Bulk Density: 3 lb/ft3 (est.) o Store: Dry, dark, cool
SPC Oil Sorbent	**Qwik-Wick**	**F-12 Fuel Pads**
o Sorbent Products Co., Inc. Maplewood, NJ 201-762-4705 o Cost: Particulate - $3.18/lb Mat - $3.50/lb Boom - $4.11/lb o Bulk Density: 4.5 lb/ft3 o Store: Dry, dark, cool	o Clark-Cutler-McDermott Co. Franklin, MS 617-528-1200 o Cost: Mats - $3.18/lb o Bulk Density: 4.4 lb/ft3 o Store: Dry, dark, cool	o Spill Control Co. Chino, CA 714-591-4704 o Cost: Mats+ - $8.60/lb o Bulk Density: 4.7 lb/ft3 o Store: Dry, dark, cool +Polymeric Foam

LIMITATIONS

Landspill

Effectiveness of particulate reduced when windy; particulate not for use within environmentally sensitive areas; pillows and mats not effective where groundcover is dense and terrain is rugged

Floating Spill - Non-Flowing Mode

None

Floating Spill - Flowing Mode

None

INCOMPATIBILITIES

o Acidic compounds, inorganic

HEALTH/SAFETY CONSIDERATIONS

Wear protective equipment appropriate for the hazardous liquid

ENVIRONMENTAL CONSIDERATIONS

Not rapidly biodegraded

SORBENT DENSITY (NO VOIDS)*

56.8 pounds/cubic foot

PERFORMANCE

Sorption capacity data and hazardous liquid/water preference index information are given in Tables 2-1 and 2-2, respectively. If the sorbent/hazardous liquid pair is either known, suspected, or observed to be incompatible, an "I" is given in Table 2-1, or if the sorbent is not wet by the hazardous liquid, an "NW" is used.

*This is the density of the solid material without any voids that is used in selected formulae contained in Section 3 and is not to be confused with sorbent bulk density listed above under specific product information.

SORBENT DATA SHEET

GENERIC CLASS: Cross-Linked Polymer

Imbiber Beads o EMCO, Inc. Little Rock, AK 501-374-7878 o Cost: Particulate - \$9.10/lb Pillows - \$11.20/lb o Bulk Density: 38 lb/ft3 o Store: Dry, cool		

LIMITATIONS

Landspill

Not reusable, effectiveness of particulate reduced when windy; particulate not for use within environmentally sensitive areas; pillows not effective where terrain is rugged or groundcover is dense

Floating Spill - Non-Flowing Mode

Not reusable

Floating Spill - Flowing Mode

Not for use on floating spills into streams or rivers - boom not available

INCOMPATIBILITIES

o Acidic compounds, inorganic

HEALTH/SAFETY CONSIDERATIONS

Wear dust respirator, eye protection and other protective equipment appropriate for the hazardous liquid. Unused beads can be a slip hazard when present on smooth, hard surface

ENVIRONMENTAL CONSIDERATIONS

Not rapidly biodegraded

SORBENT DENSITY (NO VOIDS)*

59.9 pounds/cubic foot

PERFORMANCE

Sorption capacity data and hazardous liquid/water preference index information are given in Tables 2-1 and 2-2, respectively. If the sorbent/hazardous liquid pair is either known, suspected, or observed to be incompatible, an "I" is given in Table 2-1, or if the sorbent is not wet by the hazardous liquid, an "NW" is used.

*This is the density of the solid material without any voids that is used in selected formulae contained in Section 3 and is not to be confused with sorbent bulk density listed above under specific product information.

SORBENT DATA SHEET

GENERIC CLASS: Feathers

Sea Clean Oil Sorbent o SeaClean, Inc. Miami, FL 305-661-1024 o Cost: Pillows - $7.50/lb Booms - $7.50/lb o Bulk Density: 0.89 lb/ft^3 o Store: Dry		

LIMITATIONS

Landspill

Not effective where terrain is rugged or groundcover is dense

Floating Spill - Non-Flowing Mode

None

Floating Spill - Flowing Mode

None

INCOMPATIBILITIES

HEALTH/SAFETY CONSIDERATIONS

Wear protective equipment appropriate for the hazardous liquid

ENVIRONMENTAL CONSIDERATIONS

Biodegradable

SORBENT DENSITY (NO VOIDS)*

81 pounds/cubic foot

PERFORMANCE

Sorption capacity data and hazardous liquid/water preference index information are given in Tables 2-1 and 2-2, respectively. If the sorbent/hazardous liquid pair is either known, suspected, or observed to be incompatible, an "I" is given in Table 2-1, or if the sorbent is not wet by the hazardous liquid, an "NW" is used.

*This is the density of the solid material without any voids that is used in selected formulae contained in Section 3 and is not to be confused with sorbent bulk density listed above under specific product information.

SORBENT DATA SHEET

GENERIC CLASS: Treated Clay/Treated Natural Organic

Locksorb/Klensorb o Radecca, Inc. Austin, TX 512-454-4797 o Cost: Particulate - $1.50-$5.25/lb o Bulk Density: 45 lb/ft3 o Store: Dry		

LIMITATIONS

Landspill

Not reusable or incinerable

Floating Spill - Non-Flowing Mode

Not for use on floating spills into ponds or lakes - mats and pillows not available

Floating Spill - Flowing Mode

Not for use on floating spills into streams or rivers - boom not available

INCOMPATIBILITIES

HEALTH/SAFETY CONSIDERATIONS

Wear dust respirator, eye protection and other protective equipment appropriate for the hazardous liquid

ENVIRONMENTAL CONSIDERATIONS

Does not biodegrade but clay portion is a natural component of the environment

SORBENT DENSITY (NO VOIDS)*

112.3 pounds/cubic foot

PERFORMANCE

Sorption capacity data and hazardous liquid/water preference index information are given in Tables 2-1 and 2-2, respectively. If the sorbent/hazardous liquid pair is either known, suspected, or observed to be incompatible, an "I" is given in Table 2-1, or if the sorbent is not wet by the hazardous liquid, an "NW" is used.

*This is the density of the solid material without any voids that is used in selected formulae contained in Section 3 and is not to be confused with sorbent bulk density listed above under specific product information.

3. Technical/Logistical Information

This section contains guidance on the equipment and techniques to use for: 1) the application and collection of sorbents in a spill clean-up situation, 2) the immobilization by sorbents of hazardous liquids and wastes for the purpose of disposal in a landfill, 3) sorbent reuse, and 4) the disposal of used sorbents. The purpose of this section is to provide the user with the information necessary to identify the technique and equipment requirements given the user's unique set of circumstances. Once these requirements have been identified, Section 4 may be referenced to obtain an estimate of the costs associated with the response action.

SORBENT APPLICATION - LAND SPILLS AND FLOATING SPILLS

Sorbents may be applied to a spill of a hazardous substance either manually or with mechanical aid. The choice of which general method to employ, either manual or mechanical, depends on the size of the spill and on the medium into which the spill has occurred. For example, the application of a sorbent to a small spill of 500 gallons or less of a hazardous substance onto land is generally most easily accomplished manually, using a shovel in the case of a particulate sorbent, or by throwing on a pillow, pad, or mat sorbent. On the other hand, the same size spill into a large body of water (assuming the substance floats) would probably require using a boat as a platform from which to manually apply either a pillow, pad, mat, or boom type sorbent.

As with any operation that involves a hazardous substance, there must be adequate supervision by experienced, skilled personnel who are aware of the hazards involved and the steps that can be taken to mitigate them, and who enforce strict compliance with safety procedures. The subsections that follow discuss equipment and manpower requirements, limitations, and safety precautions for the application, both manually and mechanically, of the three types of sorbents (particulate, pillow/pad/mat, boom). These subsections are limited to spills on land and spills floating on water. Technical guidance for the collection of sorbents is provided in a later subsection.

Particulate Sorbents

Manual application of particulate sorbents is limited to small spills of hazardous substances on land. Because of the difficulty in collection, particulate sorbents are not considered for use on spills into water. Also, manual application of particulate sorbents is infeasible for a large land spill because of the labor intensive nature of such an operation.

Because particulate sorbents are powders, granules, or other loose forms, the only practical method for manually applying them is shoveling. The type of shovel that is most satisfactory is a coal type scoop shovel, or other similar type with a flat bottom and turned up sides. This configuration will produce a maximum rate of application. Also, sparkproof tools should be used around flammable hazardous substances.

Particulate sorbent should be emptied from its container onto the ground upwind from the spill, and as close to it as practical and safe. Shovelfulls of the sorbent should then be applied to the spill in even, thin layers of sorbent until the top layer of sorbent is no longer wet with the spilled liquid. The personnel who apply the sorbents must be capable of skillful operation of required safety equipment for the hazardous substance involved and must observe proper safety precautions.

The volume of sorbent that can be applied per shovelfull depends on the bulk density of the sorbent. The volumetric rate of application will be inversely related to the sorbent bulk density. As an example of rates of manual application, it would require two workers approximately 40 minutes to manually apply sufficient sorbent clay to sorb a 500 gallon (4150 pounds) spill. This estimate assumes that (1) each worker could apply 50 pounds of sorbent clay to the spill each minute, (2) the sorbent clay has a sorption capacity of 1, on a weight basis, and (3) the spilled hazardous liquid has a specific gravity of one. Knowing the sorption capacity of a sorbent (Table 2-1, Section 2), its bulk density (see Sorbent Data Sheets, Section 2), and the specific gravity of the spilled hazardous liquid (Table 8-1, Section 8), the user can calculate the approximate time required for application of any sorbent to a known volume of hazardous liquid using the formula (2):

$$\text{Application Time (min)} = \frac{\text{Spill weight (lb)}}{\text{Sorption capacity (lb liquid/lb sorbent)} \times \text{Sorbent application rate (lb/min)}} \qquad (2)$$

Mechanized application of particulate sorbents to a land spill may be appropriate when the spill is of medium size (500-5000 gallons). A mechanized application technique allows for a more rapid cleanup of the spill, and makes application of the sorbent onto the center of the pool easier and more complete. When using mechanical means to apply sorbent materials to a spill, the goal should be to apply an even, uniform layer of sorbent on top of the spill from the upwind side of the spill, working from the edges towards the middle of the spill. Sorbent should be applied until the sorbent is no longer wet by the spilled hazardous substance. If the spill is into water and the hazardous liquid floats, then pillow, pad, mat, or boom type sorbents should be used since their collection is easier.

The mechanical application devices described herein have not been specifically designed to apply sorbent materials to spills; they are devices that are engineered to deliver a specific type of particulate material for a given application, (e.g., loose insulation delivered by an insulation blower). There has been no testing of these devices with any of the sorbents addressed in this Manual. They have been considered for use based solely on the general similarity of the particulate material for which they were designed to various particulate sorbent materials. Also sparkproof tools should be used around flammable hazardous substances.

The best mechanical techniques for applying particulate sorbents will depend to a large extent on the bulk density of the sorbent. For sorbents of low to moderate bulk density (e.g., cellulose-based sorbents, expanded minerals), devices that entrain the solid in a high velocity stream of air are judged to probably be the most effective. The sorbent is propelled by repeated collisions with the high-velocity air molecules. The distance the sorbent can be dispersed is limited, albeit greater than can be achieved manually, due to the comparatively low momentum which results from low density particulates.

Devices that could be used to apply low to moderate bulk density particulate sorbents are forage blowers, mulch blowers, or insulation blowers. These devices may be rented from rental yards or purchased from hardware and agricultural suppliers. The sorbent can usually be emptied into the feed hopper directly from its packaging. Sandblasting equipment is not recommended because it is designed to deliver only a small amount of material per unit time (Ref. 1).

For higher density particulate sorbents (e.g., clay minerals), centrifugal devices would probably be more effective than the air-entrainment devices described above. These devices impart momentum to the sorbent particles via a rotating impeller. A snowblower is an example of such a device that may be effective with the higher density particulate sorbents. Snowblowers may be rented or purchased at hardware stores, or large department stores, however, they may not be off-the-shelf items in the warmer areas of the country. When using a snowblower to deliver a particulate sorbent, the sorbent should first be emptied from its packaging onto the ground upwind from the spill and in a long row, one to two feet wide and eight to twelve inches high. The operator then adjusts the discharge chute for the appropriate area and pushes the snowblower into the long row of sorbent from end to end.

The mechanical devices described above each have unique methods of operation, and the instructions that are supplied with a device should be carefully consulted and followed. The normal safety precautions that are required when working around mechanical equipment, such as avoiding loose clothing and protecting hair that could get tangled in rotating parts, and using protective eyewear that safeguards against impact hazards should be observed. Safety precautions against the hazards associated with the spilled hazardous substance must also be taken. Any mechanical equipment used around a flammable hazardous substance must be explosion proof.

As an example of rates of mechanized application, it would require approximately 110 minutes to mechanically apply sufficient sorbent clay to cleanup a 5000 gallon (41500 pound) spill. This estimate assumes that (1) the application device can deliver 380 pounds of sorbent clay every minute, (2) the sorbent clay has a sorption capacity of 1, on a weight basis, and (3) the spilled hazardous liquid has a specific gravity of one. Knowing the sorption capacity of a sorbent (Table 2-1, Section 2), its bulk density (see Sorbent Data Sheets, Section 2), and the specific gravity of the spilled hazardous liquid (Table 8-1, Section 8), the user can calculate the approximate time required for application of any sorbent to a hazardous liquid using Formula 2 (given on page 3-2).

Pillow, Pad, and Mat Sorbents

Pillows, pads, and mats may be applied to spills on land or to floating spills on non-flowing water. The only practical method of application is to manually toss the sorbent onto the spill by hand, regardless of spill size or the size of the water body.

For small spills on land, pillow, pad, or mat sorbents can be applied by throwing the sorbent onto the spill either underhand or in a frisbee-throwing type of movement. This operation should be performed from the upwind side of the spill. The entire spill area should be covered with the sorbent.

For large spills on land it may be difficult to throw the sorbent onto the middle of the spill. Use a pillow type sorbent for this purpose since these are heavier than pads or mats and can therefore be thrown the farthest and are least affected by wind. If the spill is so large that the middle of the spill cannot be reached by throwing sorbent from the periphery of the spill, then proceed by working from the perimeter inwards. Place sorbent around the perimeter of the spill, then collect it and slowly work inwards toward the center of the spill.

Pillows, pads, or mats can be thrown onto a floating spill from the shore when the water body is small enough. An underhand or frisbee-throwing style will work best, and sorbent pillows will travel the farthest. If there is sufficient breeze to move the floating spill in the direction of the breeze, then one approach would be to place the sorbent around the downwind edge of the water body which will result in the hazardous substance blowing into the sorbent. Care must be exercised, however, so that placement of the sorbent does not require personnel to be stationed downwind of the spill in the vapor plume. Another, perhaps less hazardous, approach would be to place the sorbent at the upwind edge of the water body thus allowing the wind to carry the sorbent into the hazardous substance.

When the water body is large, a boat will be required so that complete sorbent application can be achieved. However, this procedure must be very carefully evaluated with respect to the nature of the spilled substance and the hazards it presents since the personnel in the boat cannot easily escape if a problem arises. A worst case scenario would be the accidental ignition of a flammable hazardous substance while the boat and personnel are near the spill, or the failure of a breathing apparatus in a spill of a hazardous vapor that is highly toxic. The only time such an operation should be attempted is if the safety of the personnel involved can be assured.

If such an operation is carried out, the sorbent should be applied as much upwind as possible by throwing the sorbent onto the spill. There should be a boat operator and a person responsible for broadcasting the sorbent. Two broadcasters per boat may be counterproductive if the boat is small since there is less room in the boat for sorbent and more frequent trips to shore will be required to resupply sorbent. This should be evaluated by the person in charge of the clean-up operation.

All safety precautions must be observed, including use of proper protective clothing. It is recommended that a line be attached to the boat, as far as is practical, so that the boat can be hauled in from shore in the event of a problem. In addition, oars are usually provided and should be checked before operations begin. All equipment must be explosion proof when working near flammable hazardous substances.

An estimate of the amount of time required to apply sufficient pads, mats or pillows to clean up a spill on water can be made using Formula 2 (see page 3-2). Sorption capacity can be found in Table 2-1, Section 2. The specific gravity data in Table 8-1, Section 8, can be used to determine the weight of a spill of known volume. The sorbent application rate can be estimated by weighing an individual pad, mat or pillow and assuming a person can throw about eight to ten of these onto the spill every minute.

Sorbent Booms

Sorbent booms (rather than pads, mats, or pillows) may be deployed against floating spills of hazardous substances in flowing watercourses when the current is below 2 knots and spill thickness is less than 1 mm. If the spill is ongoing, a sorbent boom can be deployed just downstream of the point of entry of the hazardous substance into the watercourse, or as close as practical. Additional sorbent booms could be deployed at suitable intervals downstream to collect any of the hazardous substance that has already spilled or that might subsequently escape the first sorbent boom deployed. Although sorbent/hazardous liquid compatibility was evaluated in developing the Sorbent Selection and Use Guides, the compatibility of the outer sorbent boom containment fabric with specific hazardous liquids must be verified before use.

If the spill has been halted, a sorbent boom should be placed far enough downstream so that it is completely deployed prior to the arrival of the spill slug. Backup sorbent booms should also be placed downstream to catch any hazardous substance which escapes the first-line sorbent boom.

If the spill is into a narrow stream, the sorbent boom can be deployed by throwing or casting an attached line to a person standing on the other side. The sorbent boom is then pulled across the stream. Alternatively, the sorbent boom can be hand carried across the stream -- but only if it is safe to do so.

When the spill is into a wide river, the only practical way of deploying a sorbent boom is to use a boat to carry it out from shore a sufficient distance to intercept the floating spill. Anchors may be required to secure the sorbent boom to the bottom to prevent it from being carried downstream.

Alternatively, the sorbent boom can be secured to the shoreline. The "no-oil-loss tow speed" for three sorbent booms tested was 1 ft/sec in calm water conditions (Ref. 2). Sorbent booms used for hazardous substance pickup, therefore, would be generally effective unless current flow appreciably exceeds 1 ft/sec.

The people involved in the deployment of sorbent booms must necessarily position themselves downstream of the spill. There may be times when downstream is also downwind. When such a condition exists, extra caution must be exercised so that individuals are not exposed to toxic vapors in excess of safe levels. Such precaution might include the use of fully encapsulating suits that contain SCBA (self-contained breathing apparatus). The user should refer to sources that will provide information regarding the proper protective equipment. In addition, monitoring of spill vapor concentrations may be necessary.

It is difficult to provide an estimate of the time necessary to deploy a sorbent boom against a certain size of spill because of the wide variety of circumstances that could exist in the field. However, an estimate of the amount of sorbent boom required to collect a spill can be made using the formula (3):

$$\text{Amount of sorbent boom (lb)} = \frac{\text{Spill weight (lb)}}{\text{Sorption capacity (lb liquid/lb boom)}} \qquad (3)$$

Sorption capacity can be found in Table 2-1, Section 2. The liquid specific gravity data in Table 8-1, Section 8, can be used to determine the weight of a spill of known volume. Note that Formula 3 specifies the weight of sorbent that must actually come in contact with the spill. The assumption is made that the liquid penetrates the sorbent until the sorbent is saturated. Boom effectiveness may be increased by having several booms parallel to one another (the second about 50 feet downstream of the first, etc.) and by placing fresh sorbent pads just upstream of each boom.

SORBENT COLLECTION - LAND SPILLS AND FLOATING SPILLS

Saturated sorbents may be collected or retrieved from a spill site either manually or with mechanical aid. The choice of which general method to employ, either manual or mechanical, depends on the volume of saturated sorbent that needs to be collected. Generally, when the spill has occurred on land and is 500 gallons or less, the saturated sorbent is most easily collected manually. Larger spills may warrant the use of mechanized collection equipment.

The choice of manual or mechanical collection methods for spills of floating hazardous substances into water depends more on the size of the water body than the spill size. Collection of sorbent applied to a spill that has occurred into a small body of water (e.g., a stream or pond) can probably be collected from shore manually. However, collection of a sorbent that has been applied to a spill into a large water body (e.g., a river or lake) must generally be conducted with the aid of a boat.

As with any operation that involves a hazardous substance, there must be adequate supervision by experienced, skilled personnel who are aware of the hazards involved and the steps that can be taken to mitigate them, and who enforce strict compliance with safety procedures. The subsections that follow discuss equipment and manpower requirements, limitations, and safety precautions for the collection, both manually and mechanically, of the three sorbent types for spills on land, and floating spills into water.

Particulate Sorbents

The use of particulate sorbents on spills of liquid hazardous substances is limited to land spills. Manual collection of these sorbents is recommended only for small-sized land spills because of the labor-intensive nature of such an operation. Although collection of loaded particulate sorbents can be performed by unskilled laborers, the personnel must be informed of the hazards involved and proper safety precautions to be observed.

Manual collection is best accomplished using a coal-type scoop shovel, or other similar types with a flat bottom and turned up sides. This configuration will produce a maximum rate of collection. If the hazardous liquid is flammable, all tools must be sparkproof. The workers who are collecting the saturated sorbent should be positioned on the upwind side of the sorbed spill and begin removing the saturated sorbent starting at the near edge and working toward the far edge. Ideally, the receptacle that will contain the saturated sorbent should be positioned nearby so that the saturated sorbent can be shovelled into it directly without unnecessary effort for the workers. This will also lessen the amount of dripping from the saturated sorbent that might contact unprotected ground.

The volume of saturated sorbent that can be collected per shovelfull depends on the strength of the person shovelling and therefore indirectly on the density of the sorbent and the specific gravity of the hazardous liquid. The volumetric collection rate will be inversely related to these two values. As an example of rates of manual collection, it would require approximately 80 minutes for two workers to manually collect saturated sorbent clay used to absorb a 500 gallon (4150 pounds) spill. This estimate assumes that each worker could shovel 50 pounds of saturated sorbent clay per minute, based on the estimated physical ability of an average adult male. The sorbent clay has a sorption capacity of 1, on a weight basis, and the spilled hazardous liquid has a specific gravity of one. Knowing the sorption capacity of a sorbent (Table 2-1, Section 2) and the specific gravity of the spilled hazardous liquid (Table 8-1, Section 8) allows the user to estimate the time required to collect a saturated sorbent using the formula (4):

$$\underset{\text{(min)}}{\text{Time to Collect}} = \frac{\underset{\text{(lb)}}{\text{Spill Weight}} + \underset{\text{(lb)}}{\text{Sorbent Weight}}}{\text{Mass Collection Rate (lb/min)}} \qquad (4)$$

where:

$$\underset{(lb)}{\text{Spill Weight}} = \underset{(ft^3)}{\text{Spill Vol.}} \times \underset{(lb/ft^3)}{\text{Liquid Density}}$$

and where:

$$\underset{(lb)}{\text{Sorbent Weight}} = \frac{\text{Spill Weight (lb)}}{\text{Sorption Capacity (lb liquid/lb sorbent)}}$$

A large amount of saturated particulate sorbent is most efficiently collected by mechanized means. However, the terrain must be firm and relatively flat. In addition, use of heavy equipment around flammable hazardous substances should be avoided. In such a case, manual collection of the saturated sorbent should be considered. A mechanical method of collection suitable in a wide variety of circumstances is the use of a skiploader to scrape up the saturated sorbent from the ground. The scoop of an average-sized skiploader ranges in volume from 0.5 to 1.5 cubic yards.

The skiploader should begin the collection operations upwind of the spill. Saturated sorbent should first be removed from the upwind edge, with successive passes made until the far edge of the sorbed spill is reached. The receptacle into which the saturated sorbent is to be placed should be near the collection operation to minimize the collection time. If the saturated sorbent is to be placed into 55-gallon drums, a hopper should be placed over the drum so that the skiploader can easily dump its scoopfull of saturated sorbent. Since a 55-gallon drum holds about 7 cubic feet, a scoopfull of saturated sorbent may have to be discharged into two or more separate drums.

The use of a skiploader to collect a large amount of saturated sorbent requires a skilled operator. He must be informed of the hazards involved, and proper safety precautions must be observed. As with any operation that involves a hazardous substance, there must be adequate supervision by experienced, skilled personnel who are aware of the hazards involved and the steps that can be taken to mitigate them, and who enforce strict compliance with safety procedures. The skiploader may require decontamination following use.

As an example of mechanized rates of collection, it would require an average sized skiploader approximately 40 minutes to collect saturated sorbent clay used to sorb a 5000 gallon spill. This estimated total collection time is based upon assuming (1) a sorption capacity of 1, on a weight basis, (2) a skiploader scoop capacity of 0.5 cubic yards (13.5 cubic feet), and (3) a volumetric collection rate of 1 cubic yard per minute. The user can estimate the time required to collect a saturated sorbent by the formula (5):

$$\underset{(min)}{\text{Time to collect}} = \frac{\text{Saturated Sorbent Volume (cubic feet)}}{\text{Volumetric Sorbent Collection Rate (cubic feet/min)}} \qquad (5)$$

where:

$$\text{Saturated Sorbent Volume (cubic feet)} = \text{Spill Volume (cubic feet)} + \frac{\text{Spill Weight (lb)}}{\text{Sorption Capacity (lb liquid/lb sorbent)} \times \text{Sorbent Density (lb/cubic foot)}}$$

The liquid specific gravity data in Table 8-1, Section 8, will allow the user to calculate the spill volume when the spill weight is known. Sorption capacity data are given in Table 2-1, Section 2. Sorbent density (no voids) is provided in the Sorbent Data Sheets, Section 2, and should not be confused with sorbent bulk density, also given in the Sorbent Data Sheets.

Pillows, Pads, and Mats

When a spill of small size on land has been sorbed with pillow, pad, or mat sorbents, it is generally easiest to collect the saturated sorbent manually. Manual collection is usually best accomplished with a wide pitchfork. If the hazardous liquid is flammable, all tools should be sparkproof. Using a pitchfork to collect saturated pillows, pads, or mats offers several advantages over a shovel. A pitchfork is lighter than a shovel and the saturated sorbent can either be scooped up or pierced with the tines of the pitchfork. This same technique can also be used to collect saturated mats or pillows that have been used to sorb a floating spill that has occurred in a non-flowing body of water.

When manually collecting the saturated pillows, pads, or mats from a small land spill, the workers should be positioned on the upwind side of the sorbed spill and begin removing the saturated sorbent starting at the edge and working toward the far edge. The receptacle that will contain the saturated sorbent should be positioned nearby so that the saturated sorbent can be placed into it directly, without undue effort or lengthy transport of saturated sorbent.

When manually collecting saturated pillows, pads, or mats from a small pond, it may be difficult to collect the sorbent from the upwind shore since the sorbent will be blown toward the downwind shore. However, if safe, the workers doing the collecting can don waders and enter the water from the upwind shore and collect the saturated sorbent or, alternatively, a boat can be used. These techniques will enable the workers to collect the saturated sorbent from an upwind position. If neither of these techniques are feasible, the saturated sorbent will have to be collected either with long poles or other grappling devices that can reach from the upwind shore to the "logjam" of sorbent pillows, pads, or mats, or they will have to be collected from the downwind shore. The latter method should only be attempted when adequate protection is provided. The receptacle into which the saturated sorbent is to be placed should be located near the workers to avoid any unnecessary manual transport of the saturated sorbent.

As an example of a manual collection rate, it would require approximately 30 minutes for a person to collect saturated foamed glass pillows used to sorb a 500 gallon spill. This estimate assumes that (1) the person could collect five pillows per minute, based on the estimated physical ability of an average adult male, (2) the foamed glass has a sorption capacity of 10, on a weight basis, and (3) the hazardous liquid has a specific gravity of one. Knowing the volume of saturated sorbent to be collected, the volume of a single sorbent unit (by measurement or estimate), and the rate of loaded sorbent collection, the user can estimate the time required to collect a saturated sorbent using Formula 5 (see page 3-9), where the volumetric sorbent collection rate equals the sorbent unit volume, in cubic feet/unit, times the unit collection rate, in units/minute.

Collection of saturated pillows, pads, or mats from larger bodies of water, such as lakes, will generally require the use of a boat. However, the use of a boat to collect saturated sorbent units from a lake must be very carefully evaluated with respect to the nature of the spilled hazardous substance and the hazards it presents since the personnel in the boat cannot easily escape if a problem arises. The only time such an operation should be attempted is if the safety of the personnel involved can be assured. Although skimmers have been developed and are commercially available for oil pickup, further testing of skimmers is required before skimmers can be considered for use with hazardous liquids. The relatively low viscosity of many hazardous liquids, the incompatibility of hazardous liquids with skimmer materials of construction, and the time required for deployment are obstacles that limit the utility of skimmers for hazardous liquid cleanup (Ref. 3).

If it is determined that it is safe to utilize a boat for the collection operation, the crew should consist of 1) a skilled boat operator who maintains the boat upwind of the saturated sorbent and maneuvers it close enough to the floating saturated sorbent units to allow for easy retrieval and 2) one or more workers positioned forward who gather up the saturated sorbent with pitchforks or other suitable implements. The boat should be sufficiently large to allow for easy maneuvering by the crew and with space for drums into which the collected saturated sorbent can be placed. The use of motor boats and other potential fire or explosion initiators should be eliminated when collecting saturated sorbent at the site of a spill of a flammable hazardous substance. Also, the boat should be tethered, as far as is practical, to shore so that it can be hauled in if the crew is incapacitated for any reason.

An estimate of the time required to collect the saturated sorbent can be made by using Formula 5 (see page 3-9), where the volumetric sorbent collection rate equals the sorbent unit volume, in cubic feet/unit, times the unit collection rate, in units/minute. (Approximately 6 floating mats or pillows collected per minute from a boat is estimated to be a reasonable unit collection rate.)

Sorbent Booms

The use of sorbent booms on spills of floating liquid hazardous substances is best for spills into flowing bodies of water (i.e., creeks or rivers), although booms are also useful for protecting windward shores of quiescent bodies from spills. When the spill is into a small creek, the sorbent boom can usually be manually collected. This operation should be conducted from shore and upwind, if possible. The boom is pulled from the water and lifted directly into a suitable receptacle. This receptacle (e.g., a 55-gallon drum) should be placed as near to the shore as possible to make collection easier and to minimize the amount of hazardous substance that drips from the sorbent boom. The personnel involved in the manual collection operation must be suitably protected against the hazardous substance, particularly since handling of the saturated sorbent boom may be required.

When the spill is into a river, mechanical means of collection include boats and winches. When the saturated sorbent boom is not connected to the same shore, a boat will be required for collection. Otherwise, the saturated sorbent boom can be hauled in from shore. In either case, a winch may be required to hoist the sorbent boom out of the water since it can be heavy as well as lengthy. However, care should be exercised when using a winch to avoid excessive squeezing of the sorbent and resultant liberation of the hazardous liquid. Unless the sorbent boom is to be reused (upon liquid release into a suitable receptacle by squeezing), the collected sorbent boom should be placed directly into a suitable receptacle (e.g., a 55-gallon drum), which should be placed as close to the point of collection as possible.

All collection operations, whether on shore or from a boat, should take place from an upwind position. In addition, the use of a boat must be evaluated in light of the nature of the hazardous liquid and the dangers involved. The same precautions and requirements for the utilization of a boat during sorbent collection that were described for pillow, pad or mat sorbents also apply to collection of saturated sorbent booms.

An estimate of the time required to collect a saturated sorbent boom can be made by using Formula 5 (see page 3-9), where the volumetric sorbent collection rate equals the sorbent unit volume, in cubic feet/unit, times the unit collection rate, in units/minute. (A single saturated sorbent boom of 10 foot length collected in about 5 minutes is estimated to be a reasonable unit collection rate.)

SORBENTS FOR LANDFILLING OF LIQUIDS

Landfills that accept containerized hazardous waste for disposal are required to comply with EPA regulations (47 FR 32366, July 26, 1982) regarding the disposal of those containers. Containers holding liquids or free liquids must not be placed in a landfill unless, before placement, the liquids or free liquids are treated or stabilized chemically or physically (e.g., by mixing with a sorbent solid), so that free liquids are no longer present. This section discusses techniques that can be used to accomplish physical stabilization via sorbents and provides the user with a way to estimate the volume of liquid that can be properly and safely placed in each 55-gallon drum.

Clean, dry drums, constructed of materials compatible with the liquid hazardous substance to be disposed of, and in compliance with all federal, state, and local regulations should be used. Sufficient dry sorbent should be added to the drum to make a layer 6 to 8 inches deep. Next, a sufficient amount of the hazardous liquid is added to the drum to cover the sorbent layer. This sorbent/liquid mixture should then be stirred with a rod made of compatible material to assure complete wetting of the sorbent. These steps should be repeated until the saturated sorbent is about 6 inches from the top of the drum. Add another couple of inches of dry sorbent to provide an excess sorption capacity should the sorbent settle during transport to the landfill. Allowance for ullage should be based on estimated liquid volume, anticipated exposure temperatures, and the liquid expansion coefficient.

The volume of hazardous liquid that can be disposed of in a drum can be estimated using the following formula (6):

$$V_{HL}/\text{drum} = \frac{V_{drum} - V_{ullage}}{1 + [\,(D_{HL} \times 62.4) / (SC \times D_s)]} \qquad (6)$$

where: V_{HL} = Volume of hazardous liquid (gal)
V_{drum} = Volume of the drum (gal)
V_{ullage} = Ullage volume (gal)
D_{HL} = Density of the hazardous liquid (gm/cc)
SC = Sorption capacity (lb HL/ lb dry Sorbent)
D_s = Density of the sorbent (no voids) (lb/cu ft)

The liquid specific gravity (density) is available in Table 8-1, Section 8. Sorption capacity is provided in Table 2-1, Section 2. The sorbent density is given in the appropriate Sorbent Data Sheet, Section 2, and should not be confused with sorbent bulk density, also given in the Sorbent Data Sheets.

SORBENT REUSE

Many sorbents have the necessary compressibility, hysteresis, and durability that allow them to be reused a number of times during the course of a spill cleanup operation. Sorbent reuse is indicated when recovery of the spilled hazardous substance in free liquid form is desired. The sorbent functions solely to pickup the free liquid from the ground and to release the sorbed liquid into a recovery vessel.

Recovery of a spilled hazardous substance in free liquid form may be desired when 1) the value of the substance is great enough that recovery and recycle is less costly than the combined cost of disposal and loss of product, or 2) the recovered hazardous substance is to be incinerated which does not require the liquid to be solidified or otherwise immobilized. The cost/benefit of the latter case, in particular, must be carefully analyzed. Benefits include the need for less sorbent and a smaller volume of material that requires incineration with concomitant savings on drum and transportation costs. The primary costs are increased manpower per unit of liquid sorbed and the cost of recovery equipment.

When analysis indicates that sorbent reuse is preferable to single-pass application, the user should pay particular attention to the "Limitations" column in the Selection and Use Guide. Here, "R" indicates that a sorbent is not reusable. When an "R" does not appear, the sorbent can be reused. Generally, pillows, pads, and mats are the best forms to employ when reuse is to be attempted. (While "R" does not appear in the "Limitations" column for particulate sorbents that are compressible and exhibit hysteresis, it is suggested that particulate forms only be reused when specially designed equipment is available for handling and squeezing such sorbent (Ref. 5).

To reuse a sorbent, a device must be used that applies a squeezing action to the sorbent to release the sorbed liquid into a receptacle (e.g., a 55-gallon drum). Such a device could be fabricated or various wringers can be purchased from hardware stores, supply companies, and many sorbent suppliers. These can either be manually operated or powered by electric motors. A powered wringer must be explosion proof when used with flammable hazardous substances. Wringing with one's hands is not recommended because of possible contact with the hazardous substance and because the twisting action may damage the sorbent.

SORBENT DISPOSAL

Used sorbents can be disposed of by landfilling or incineration (in either case, the loaded sorbent is generally contained in 55-gallon drums for shipment to the disposal or treatment site). The decision of which disposal method to use will be made on the basis of cost and the disposal requirements of the landfill or incinerator operator.

Incineration of used sorbent is a destructive process that uses heat and oxygen to oxidize the hazardous substance to innocuous or less noxious compounds. Ideally, the sorbent should be combustible so that a minimum of residue is produced during incineration. If it is known prior to cleanup that incineration is the method of choice for disposal and that the incinerator can accept solid waste, combustible sorbent should be chosen. This can be done by avoiding those sorbents listed in the Selection and Use Guide that show an "I" in the "Limitations" column. These sorbents are not combustible. In addition, an analysis should be made to determine the feasibility of reusing sorbent during cleanup since incineration of free liquids is often preferable to incineration of solids. (See Sorbent Reuse above in this section.)

Landfilling of used sorbent is subject to a number of federal, state, and local regulations regarding what type and form of wastes can be placed in a landfill. The details of these requirements should be carefully studied if landfilling of the used sorbent is being considered. A telephone call to the landfill operators may be the quickest way to determine the requirements. Recent regulations (47 FR 32366, July 26, 1982) prohibit the containerized disposal of any free-standing liquid in hazardous waste landfills. It, therefore, is advisable to use sufficient sorbent to assure that the hazardous liquid is completely sorbed.

4. Cost Estimation Procedures and Data

This section provides the user with price data for the purpose of estimating the cost of using a sorbent to clean up a spill of hazardous liquids or to immobilize a hazardous liquid to meet the "no free liquid" criterion for the purpose of landfilling. A discussion of the general cost categories for cleanup of a hazardous substance spill with sorbents is presented. The major categories are equipment, expendable materials, labor, disposal, and transportation. Some cost data (1983) are given for the user's cost estimation purposes and must be adjusted to current levels using appropriate cost indices. Manual Sections 2 and 3 should be consulted prior to using this section to determine specific equipment, materials, labor, disposal, and transportation requirements unique to the particular spill situation confronting the user.

EQUIPMENT

Cost data are provided for equipment used to apply and collect sorbents during the cleanup of a CERCLA liquid spill. Where available, data are given for purchase and rental of the equipment. One of the primary factors that will influence the decision to rent or buy equipment is the relative cost of decontamination versus disposal. Rented equipment must be completely decontaminated before it can be returned, therefore, the labor and decontamination material costs must be added to the rental cost to obtain the bottom line cost of using the rented piece of equipment. On the other hand, purchased equipment may either be disposed of (by an approved method) without decontamination, or decontaminated and stored for future use. The anticipated length of time the equipment will be required is another important factor that must be considered when making a rent/buy decision. Much of the cost data contained in the tables in this section were obtained from the 1983 Dodge Guide to Heavy Construction Costs (Ref. 6), and the 1983 RER (Miramar Publishing) Rental Equipment Rate Guide (Ref. 7).

Sorbent Application and Collection Equipment

In Table 4-1 in this subsection, cost data on sorbent application and collection equipment are given.

Miscellaneous Equipment

Table 4-2 shows the cost data for miscellaneous equipment required during the cleanup of a spill of hazardous liquid or during the landfilling of a sorbent containing a hazardous liquid. This equipment includes support equipment not directly involved in the cleanup, such as trailers, but does not include safety related equipment.

Table 4-1. SORBENT APPLICATION AND COLLECTION EQUIPMENT PRICE DATA

ITEM	RENTAL PRICE($)	PURCHASE PRICE($)
Shovel, Pitchfork	5/day	15-30
Large forage blower	-	11,500
Snow blower	40/day	450-1000
Grader	110-200/day	-
Skiploader	145-160/day	-
Vacuum truck	200-800/day	-
Mechanical shovel	575-1700/day	-
12ft-14ft workboat, w/o motor	15-20/day	500-700
12ft-14ft jon boat, w/o motor	15-20/day	420-900
10-15 Hp outboard motor	20/day	500-1000
30ft-40ft diesel workboat	280-450/day	-
12ft by 8ft steel barge	40/day	-
Diesel tractor	300/day	-

TABLE 4-2. MISCELLANEOUS EQUIPMENT COSTS

ITEM	RENTAL COST($)	PURCHASE COST($)
Rotating 55 gal. drum mixer	-	1000
55 gallon drum hand truck	8/day	200
Transfer pump	7/day	250
One-ton truck	48/day	-
Two-ton truck	54/day	-
Pickup truck/van/car	35-55/day	-
Decontamination trailer w/generator	350/day	-
Lowboy trailer	145/day	-
Flatbed trailer	75/day	-
Office trailer	20/day	-
Galley trailer	200/day	-
Dump truck	105/day	-

MATERIALS

Cost data are provided here for expendable materials used during sorbent cleanup of a hazardous substance spill or for sorbent landfilling purposes. Included are cost data on sorbents and on the items required for safe disposal of a sorbent saturated with a hazardous liquid. The amount of sorbent required can be calculated from the sorption capacity of the sorbent (see Table 2-1 in Section 2) and the estimated quantity of hazardous substance.

Sorbent cost ranges are contained in Table 4-3. For costs of specific sorbents, the user can refer to the Sorbent Data Sheets in Section 2, while cost data for other expendable materials are contained in Table 4-4. Volume discounts may be available and should be investigated as warranted.

TABLE 4-3. SORBENT PRICES

GENERIC SORBENT CLASS	PRICE ($/LB) PARTICULATE	PILLOW, PAD, MAT	SORBENT BOOM
Wood fiber	0.10-0.21	2.00	NA
Treated wood fiber	1.75	1.48-10.00	1.55-3.75
Feathers	NA	7.50	7.50
Sorbent clay	0.04-1.50	NA	NA
Diatomite	1.45	NA	NA
Expanded mineral	0.12-0.31	NA	NA
Polypropylene	2.00-3.18	2.00-8.60	2.57-4.30
Polyethylene	NA	4.25	NA
Polyurethane	NA	6.16-15.00	7.83-15.00
Cross-linked polymer	9.10	11.20	NA
Foamed glass	3.00	3.20-4.40	NA
Treated clay/treated natural organic	1.50-5.25	NA	NA

Note: NA = Not applicable

TABLE 4-4. NON-SORBENT (EXPENDABLE MATERIAL) PRICES

ITEM	ESTIMATED PRICE($)
55 gal. steel drum	12-50
55 gal. polyethylene drum	120
Polyethylene bags	
9"x12" (4 mil)	0.07
12"x18" (4 mil)	0.12
55 gal. (3 mil)	0.45

LABOR

Labor requirements for spill cleanup will depend on the quantity and nature of hazardous liquid to be removed and the urgency with which the removal must be accomplished. The necessary labor skills are not just limited to the ability to apply and collect particulate, pad, mat, pillow or boom sorbents, but also include the ability to operate safely in proximity to hazardous substances. Instances have occurred in which safety-related problems arose when personnel performing sorbent-related cleanup tasks were not sufficiently cautious due to lack of appropriate safety training, experience, and supervision. Well trained personnel will be required to assure that proper procedures are being adhered to and to evaluate site conditions. For the purpose of estimating labor costs, salary scales (including overhead rates) for several applicable personnel skill categories are given in Table 4-5.

TABLE 4-5. LABOR CATEGORY COSTS

Labor Category	Estimated Cost for Straight Time ($/hr)
Consultant	75-110
Supervisor	15-45
Safety Representative	30
Engineer	40-45
Chemist	35-55
Biologist	35-40
Hydrologist/Geologist	45-50
Foreman	13-30
Recovery Technician	18-19
Lab technician	24
Laborer	15
Equipment Operator	20
Truck Driver	16
Courier, Typist	16
Mechanic	22
Salvage Master	50

Working with hazardous substances may also require a surcharge of up to $125/day/person, which includes special protective clothing. Labor cost estimates should also include food and lodging expenses (per diem) for personnel who are unable to return home after work. Per diem costs typically range from $50-100.

DISPOSAL COSTS

Landfilling

Landfill disposal costs of the sorbent/hazardous liquid combination will depend to a great extent on the type of material sorbed. Packaging requirements and costs will vary depending upon the sorbed material. For example, extremely toxic sorbed material may require that the sorbent/hazardous liquid be placed in drums before disposal whereas, in other cases, bulk disposal might be satisfactory. The range of costs for landfilling drummed and bulk wastes are as follows (Ref. 8):

Drum - \$168-\$240/ton
Bulk - \$55-\$83/ton

Incineration

Incineration of a sorbent which contains a hazardous liquid is limited to those incinerators that can handle solid waste, e.g., rotary kilns. The range of costs for incinerating drummed and bulk solid waste are (Ref. 8):

Drum - \$120-\$400/ton
Bulk - \$395-\$790/ton

TRANSPORTATION COSTS

Transportation costs are incurred during initial mobilization of necessary equipment, materials, and personnel for a spill response and also during transport of the loaded sorbent wastes to the disposal site.

Mobilization

Mobilization costs would consist of rental fees (see Table 4-2) and mileage charges for vehicles necessary for the cleanup operations, such as tractor/trailers, dump trucks, offroad 4-wheel drive trucks, pickup trucks, and various trailer units containing generators, galleys, offices, etc. Other possible mobilization costs might include rental costs for airplanes and/or helicopters to be used for transportation or surveillance purposes.

Mobilization costs will depend primarily upon the size of the spill and the spill location. These two factors will dictate the amount of equipment and personnel necessary to effect the cleanup and hence will determine the mobilization costs. Typical mileage charges for various types of vehicles, along with rental charges for aircraft, are given in Table 4-6.

Disposal

Transportation costs will also be incurred during transport of the loaded sorbent to the ultimate disposal site. Depending upon the nature of the spilled material (flammability, corrosivity, etc.) specialized transportation equipment may be necessary. Estimated costs for hauling the wastes are \$2-4 (40-80 drums) per loaded mile for a hazardous waste disposal truck.

TABLE 4-6. CONVEYENCE MILEAGE CHARGES

Conveyance	Mileage Charge
Automobiles	32¢/mile
Diesel tractor	56¢/mile
Pickup truck	47¢/mile
Personnel truck	47¢/mile
Dump truck	52¢/mile
Vacuum truck	58¢/mile
Aircraft	
Plane:	$100-$320/hour $350/hour with pilot $2.30 to $2.50/round-trip mile
Helicopter:	Up to 3 hours....$165-$310/hour >3 hours........$400-$575/day or..............$ 75-$95/flight hour

5. Test Methods

This section describes the test methods used to determine the experimental sorption capacity values reported in Table 2-1. Other test methods that are available for evaluating sorption capacity and related characteristics, such as, release rate and swelling, are also described. Table 5-1 lists the test methods and provides pertinent information such as the type of data generated with these test protocols and the resources required to perform the various tests. The use of standardized test methods for quantifying sorbent performance and characteristics have not been universally adopted or applied by manufacturers, except by those manufacturers wanting to qualify for bidding government procurements. However, these standard government tests of performance for the purpose of procurement are concerned with oil and waste clean-up from floor surfaces, not the removal of hazardous substances from the environment.

In an effort to address the need for a standard test method for sorbents used to sorb hazardous substances, the American Society for Testing and Materials (ASTM) recently promulgated standard F716-82; "Standard Method of Testing Sorbent Performance of Absorbents", under the jurisdiction of ASTM Committee F-20 (Hazardous Substances and Oil Spill Response) and Subcommittee F20.12 (Oil Removal). However, this standard is relatively new (1982) and few data have been generated using the procedures specified therein. Attempts to implement ASTM F716-82 for the purposes of generating data useful in the development of this Manual led to design and implementation of a new protocol for measurement of sorption capacity of various types of sorbents. The test protocol consists of (1) a decant procedure for particulate sorbents, (2) a visual volume procedure for finely-divided particulate sorbents, and (3) an immersion procedure for pad/mat sorbents. These test procedures had their origins in the ASTM F716-82 standard but were significantly modified.

The sparse data on sorption of hazardous liquids by sorbents that are available from manufacturers' marketing brochures generally are not accompanied by information concerning the test method. Manufacturers of low density sorbents tend to report sorption capacity data in terms of the mass sorbed per unit mass of sorbent. Manufacturers of the heavier sorbents, such as the natural earth minerals, often report results in terms of the mass sorbed per unit volume of sorbent. The importance of the choice of unit can be illustrated by comparing the sorption capacity of a sorbent of high bulk density to that of a sorbent of low bulk density on both a per unit weight and a per unit volume basis. For example, the sorption capacities of vermiculite and sorbent clay with acetonitrile are as follows:

TABLE 5-1. TEST METHODS FOR MEASURING SORBENT PERFORMANCE

Test Method/Source	Applicability/Limitations	Data Type	Testing Time (Hrs)	Testing Equipment
CE-EMSI Sorbent Tests ----- EMSI 2421 W. Hillcrest Newbury Park, CA 91320	o Particulate pillow, pad, or mat sorbents or sorbent booms	o sorption capacity	2 (fume hood)	graduated cylinder, wire screen assembly, glass tray, balance, spatula, beaker,
	o Various sorbent classes	o hazardous liquid/water preference ratio	-	none (calculation)
	o Floating or sinking hazardous liquids or land spills			
ASTM F716-82 ------- ASTM 1916 Race St. Philadelphia, PA 19103	o Particulate, pillow, pad, or mat sorbents or sorbent booms	o sorption capacity	2-24	graduated cylinder, stopper glass rod, screen, balance
	o Various sorbent classes	o rate of release	2 (max)	glass dish, fume hood, balance
	o Sinking hazardous liquids, spills on land	o penetration	24	glass tube, stopper sand, dye, glass bottle, 550g weight
		o release to other materials	2 (est.)	glass dish, dye, blotting paper, balance, weight
		o humidity/water resistance	24	glass jar w/lid, balance
		o swelling	2-24	graduated cylinder, stopper, glass rod, screen, balance
GSA P-A-1056B ----- General Services Administration Washington, D.C. ----- Available through NTIS 5285 Port Royal Rd. Springfield, VA 22151	o Particulate (silicate mineral) sorbents	o sieve size	2 (est.)	U.S. std. sieve series, balance
	o Oil and water mixtures	o resistance to attrition	1 (est.)	U.S. std. sieve series, sieve shaker, steel balls, balance
	o Potential hazardous liquid applications	o workmanship	1/4	----
		o aqueous solubility	48	oven balance, beaker, filtration apparatus, crucible
		o sorption capacity	6	glass tube, screen, balance

(continued)

TABLE 5-1. (continued)

Test Method/Source	Applicability/Limitations	Data Type	Testing Time (Hrs)	Testing Equipment
Westinghouse ----- Westinghouse Electric Corp., The Westinghouse Bldg. Gateway Center Pittsburgh, PA 15222	o Particulate (silicate mineral) sorbents o Oil and oil/water mixtures o Potential hazardous liquid application	o sorption capacity	3	mesh cone, beaker, balance
Rensselaer Report No. CG-D-38-76 ----- Rensselaer Institute 110 Eighth St. Troy, NY 12181	o Particulate, pillow, pad, or mat sorbents or sorbent booms o Various sorbent classes o Floating hazardous liquids or land spills	o sorption capacity	1/4-2 (est.)	glass dish, balance
USCG Report No. 724110.1/2/1 ----- United States Coast Guard Dept. of Transportation Washington D.C ----- Available through NTIS 5285 Port Royal Rd. Springfield, VA 22151	o Particulate, pillow, pad, or mat sorbents or sorbent booms o Various sorbent classes o Potential application for floating hazardous liquids or land spills	o buoyancy retention	6 (max)	glass bottle, screen basket, watch glass, balance
		o sorption capacity	6 (max)	screen basket, balance beaker, glass rod, spatula, watch glass, glass bottle stopper, shaker
		o hazardous liquid retention	-	none (calculation)
		o hazardous liquid/ water content ratio	6	glass bottle, shaker, balance
		o hazardous liquid/ sorbent coherence (particulates)	1/4	---
		o sorbent reusability	1/2	pan, wire screen, balance, wringer

(continued)

TABLE 5-1. (continued)

Test Method/Source	Applicability/Limitations	Data Type	Testing Time (Hrs)	Testing Equipment
Environment Canada	o Particulate, pillow, pad, or mat sorbents or sorbent booms	o sorption capacity	1	constant temperature bath, laboratory press, balance, assorted glassware
Report Nos. EPS-4-EC-76-5 EPS-4-EC-78-8 EPS-4-EC-83-4 ----- Technical Services Branch, Environment Canada - EPS River Road Labs Ottawa, Canada KIA IC8	o Various sorbent classes o Potential application for floating hazardous liquids or land spills	o hazardous liquid/ water content ratio	-	none (calculation)

Sorbent (Bulk Density)	Sorption Capacity with Acetonitrile	
	g/g	lb/ft^3
Vermiculite (5 lb/ft^3)	3.8	19.0
Sorbent Clay (38 lb/ft^3)	0.9	34.2

When the data are compared on a per weight basis, the vermiculite appears the better performer. However, on a per volume basis, the sorbent clay sorbs a greater mass of hazardous liquid than vermiculite.

The absence of a data base of sorbent performance information derived from common test methods makes comparison of results difficult. Descriptions of various sorbent test methods are thus presented in this section so that the user can (1) make a judgement as to the best test method to use for his unique cleanup situation, if required, or (2) understand how the experimental values reported in Table 2-1 were derived. This information on standard test methodologies is also presented to provide further insight to the user regarding the differences between laboratory test conditions and the conditions that exist in the field. This section provides a detailed description of the test methods used to generate date for this project, as well as brief descriptions of other available test methods. The descriptions include purpose, applicability/limitations, data type and resource requirements. Note: The user should obtain the detailed test method protocol from the issuing organization before attempting any actual testing (see Table 5-1 for source addresses). Also, sorption capacity testing should be performed by investigators familiar with proper laboratory procedure and using correct safety equipment, clothing, and respirators.

SORBENT TESTING PROCEDURES USED FOR THIS PROJECT

Several new standard operating procedures for measuring sorption capacities of particulate and pillow, pad, mat sorbents were developed for this project and are described in the following paragraphs.

Sorption Capacity Procedure for Particulate Sorbents

For particulate sorbents that are not finely-divided the following protocol for determining sorption capacity is conducted with an excess of fluid. In a 10-ml graduated cylinder, place 10-ml of sorbent (2-ml for cross-linked polymer) and weigh to determine the weight of sorbent. Place the 10-ml (or 2-ml) of sorbent into a 100-ml graduated cylinder that has been previously tared. Add about 50-ml of fluid to be tested, then suitably cap the mouth of the cylinder. Periodically swirl the mixture to assure intimate contact of sorbent with hazardous liquid. Monitor the sample temperature during the 2-hour test. At the conclusion of the exposure period, decant excess liquid from the cylinder by tipping or inverting the cylinder with a 200-mesh SS screen (housed in a special frame) pressed tightly against the cylinder mouth. Position a glass tray below the screen to catch the decanting liquid. When the dripping ceases (less than 1 drop/15 seconds), the cylinder within is weighed with the wet sorbent inside. The weight of hazardous liquid sorbed is the difference between the wet and dry sorbent weights. The weighings obtained from the test protocol are entered onto a Sorption Capacity Data Sheet - Decant Test. Calculations are then made as prescribed above and the sorption capacity is obtained as grams of hazardous liquid per gram of dry sorbent and as pounds of hazardous liquid per cubic foot of dry sorbent. The testing is conducted in duplicate and the spent sorbent/hazardous liquid mixture is disposed of in specially designated wide-mouth waste disposal bottles. If the results from the duplicate tests are not within 20% of each other, a third test is conducted and the average value and standard deviation reported.

Sorption Capacity Procedure for Finely-Divided Particulate Sorbents

For finely-divided particulate sorbents, the following experimental approach is used to obtain sorption capacity. In a 50-ml graduated cylinder, place approximately 25 mls of sorbent and weigh to determine the weight of sorbent. Add an excess of fluid to be tested, i.e., enough to adequately submerge the sorbent, swirl the mixture to assure intimate mixing, then suitably cap the mouth of the cylinder and allow the sorbent within to settle. Monitor the room temperature as an indication of sample temperature during the 2-hour test. At the conclusion of the exposure period, weigh the cylinder and contents, thus obtaining the weight of hazardous liquid present. Using the hazardous liquid density, calculate the volume of hazardous liquid and visually determine the total volume of sorbent and liquid. By subtracting the calculated volume of liquid, the actual volume of sorbent is obtained. Now visually measure the level of wet sorbent residing in the bottom of the graduated cylinder to obtain the observed volume of wet sorbent. By subtracting the actual volume of sorbent from the observed volume of set sorbent, the volume of liquid sorbed by the sorbent is calculated. The weighings and observations from the test procedure are entered onto a Sorption Capacity Data Sheet - Visual Volume Test. Calculations are then made as prescribed above and the sorption capacity is obtained as grams of hazardous liquid per gram of dry sorbent or as pounds of hazardous liquid per cubic foot of dry sorbent. The testing is performed in duplicate and the spent sorbent/hazardous liquid mixture is disposed of in specially designated wide-mouth waste disposal bottles. If the results from the duplicate tests are not within 20% of each other, a third test is conducted and the average value and standard deviation reported.

Sorption Capacity Procedure for Pillow, Pad, and Mat Sorbents

For pillow, pad or mat sorbents, the following protocol is applied to obtain the sorption capacity. A standard-sized (usually 15 cm on edge) piece of sorbent is weighed and then placed in contact with hazardous liquid in a glass tray. Excess liquid should be used to assure total saturation of the sorbent. Room temperature is monitored and recorded. After a 2-hour saturation, the sorbent unit (pillow, pad, or mat) is removed by a corner of the unit from the fluid in the tray. The fabricated unit must remain handleable (without tearing) at the saturation point. The sorbent unit is then hung from a special weighing assembly that includes a glass tray to catch the drippings. Upon cessation of dripping (less than 1 drop/15 seconds), the sorbent unit is weighed with the glass tray (and drippings) removed from the assembly. The weight is recorded at the time dripping stops as the maximum practical pickup amount. The weighings obtained from the test protocol are entered onto a Sorption Capacity Data Sheet - Pillow/Pad/Mat Test. Calculations are then made as prescribed above and the sorption capacity is obtained as grams of hazardous liquid retained per gram of dry sorbent or as pounds of hazardous liquid per cubic foot of dry sorbent. The spent or loaded sorbent is disposed of in specially designated wide-mouth waste disposal bottles. Makeup fluid is added as required to the tray and the test is repeated a second time with a new sorbent unit immersed in the hazardous liquid. If the two values are not within 20% of each other, a third measurement is conducted and the average value and standard deviation reported.

ASTM F716-82

The purpose of the ASTM F716-82 test method entitled, "Standard Methods of Testing Sorbent Performance of Absorbents" is to develop laboratory test data which describe the performance of sorbent materials used to remove oils and other compatible fluids from water. The method is described as being useful with particulate, pillow, pad, or mat sorbents or sorbent booms of different construction and materials. It is applicable to hazardous liquids that float or sink in water. The ASTM test method is one of the more extensive of those identified and included in this presentation. While most of the other test methods consider only sorption capacity as a measure of performance, the ASTM test method also considers rate of release, penetration, release to other materials, humidity/water resistance, and swelling. Swelling would be of particular importance when immobilizing a liquid for the purpose of landfilling. Rate of release data can be used to estimate the degree of hazard in the case of flammable or otherwise hazardous substances.

GSA P-A-1056B

The General Services Administration (GSA) test method P-A-1056B entitled, "Federal Specification, Absorbent Material, Oil and Water For Floors and Decks" is designed to develop laboratory test data which describes the performance of sorbent materials for use on floors and decks. The method may be used with particulate sorbents of silicate mineral composition. It is applicable to oil and water but can probably be applied to hazardous liquids as well. Only a single performance test is specified: sorption capacity.

WESTINGHOUSE

Like the GSA Method, the Westinghouse Test Method is designed to develop laboratory test data which describe the performance of sorbents for use on floors and decks. The method may be used with particulate sorbents of silicate mineral composition. It is applicable to oil or oily water mixtures, but can probably be applied to hazardous liquids as well. This method tests the single performance factor of sorption capacity.

RENSSELAER

The Rensselaer Test Method was developed by workers at the Rensselaer Polytechnic Institute in New York and published in the report entitled, "Agents, Methods and Devices for Amelioration of Discharges of Hazardous Chemicals on Water" (Coast Guard Report No. CG-D-38-76, published in 1975). The method may be used with particulate, pillow, pad, or mat sorbents or sorbent booms of different construction and materials. It is applicable to hazardous liquids that float in water or that are spilled on land. The single performance test measures sorption capacity.

USCG

This test method was developed at the Naval Ship Research and Development Center and published in the report entitled, "Investigation of Sorbents for Removing Oil Spills from Waters" (Coast Guard Report No. 724220.1/2/1, published in 1971). The method may be used with particulate, pillow, pad, or mat sorbents or sorbent booms of different construction and materials. Although originally intended for testing sorbents used for cleanup of oil spills on water, this method is applicable to any hazardous liquid that floats on water or is spilled on land.

ENVIRONMENT CANADA

This test method was developed for testing oil spill sorbents and is detailed in three reports issued by Environment Canada entitled, "Selection Criteria and Laboratory Evaluation of Oil Spill Sorbents" (Technology Development Report EPS-4-EC-76-5 published in 1976), "Selection Criteria and Laboratory Evaluation of Oil Spill Sorbents: An Update" (Technology Development Report EPS 4-EC-78-8, published in 1978) and "Selection Criteria and Laboratory Evaluation of Oil Spill Sorbents: Update II" (Technology Development Report EPS 4-EP-83-4, published in 1983). This test method is applicable to hazardous liquids that float or that are spilled on land. Particulate, pillow, pad, or mat sorbents or sorbent booms can be accommodated by this test.

6. Spill Scenario Rationale

This section presents the rationale that was used for choosing the three cleanup scenarios considered in this Manual: (1) spills onto land, (2) floating spills into water, and (3) disposal in landfills. All of these scenarios represent situations where sorbents may be a viable method of spill amelioration or containment for disposal. Other release scenarios, such as release into air, were rejected because of the low probability of using sorbents for amelioration of such spills. An example of a scenario which was not addressed is the spilling of a soluble hazardous substance into water, since the use of hydrophobic sorbents, used for gross cleanup of insoluble hazardous liquids into water, is relatively ineffective in these situations compared with other techniques such as carbon adsorption. In addition, carbon adsorption was defined to be beyond the scope of this Manual since considerable data already exists on the subject and carbon adsorption is usually employed to polish an effluent stream, not to effect a gross cleanup as would be the case with the sorbents considered in this Manual.

SPILL ONTO LAND

Sorbents can be quite effective for the cleanup of spills onto land, particularly when the size of the spill is small, for example, less than 500 gallons. In fact, using sorbents for small spills is often the method of choice because other amelioration techniques are too expensive on a small scale or are otherwise impractical. Unconfirmed analysis of spill size data indicates that most spills are 100 lbs or less and that most spills are onto land. In light of the predominance of small spills onto land compared with other spill sizes and media, it is appropriate to address this scenario via the Manual.

FLOATING SPILL INTO WATER

Spills of hazardous substances into water are also common occurrences, but as with spills onto land, spills into water are usually small and probably occur primarily during bulk cargo transfer operations. Small spills of floating hazardous substances into water are generally most easily cleaned up with sorbents. Other methods of cleanup are ineffective on this type of spill because they generally collect large amounts of water with the recovered hazardous substance, thus requiring another separation step.

Care must be exercised when choosing a sorbent to clean up a floating hazardous substance because there are two liquid phases involved, not a single liquid phase as with a land spill. The sorbent must selectively sorb the hazardous substance instead of water and must not sink. In other words, only hydrophobic floating sorbents are considered for this scenario. Also, particulate sorbents that float are not considered feasible since their collection would be difficult. In a non-flowing or slow moving water body, sorbent booms and pillow, pad, and mat sorbents can be effective. In a flowing water body, sorbent booms may be effective depending on the water flow and the size of the spill. Also, depending on the nature of the flowing water body and the degree of containment rendered by the sorbent boom, pillow, pad, or mat sorbents may be used effectively in a complementary role.

LANDFILLING

Landfills that accept containerized hazardous waste for disposal are required to comply with EPA regulations (47 FR 32366, July 26, 1982) regarding the disposal of those containers. Containers holding liquids or free liquids must not be placed in a landfill unless, before placement, the liquids or free liquids are treated or stabilized chemically or physically (e.g., by mixing with a solid sorbent), so that free liquids are no longer present. Free liquids can thus be retained by a sorbent or otherwise immobilized. This scenario is therefore addressed in the Manual so that a source of information will be available to guide the user in the proper choice of sorbent for liquid hazardous substance disposal.

7. Sources of Information

To develop the Sorbent Selection and Use Guides and the Sorbent Data Sheets, it was necessary to obtain or develop relevant data on both liquid hazardous substances and sorbents. Data on CERCLA liquids were required to identify the CERCLA hazardous liquids, their disposition upon release in the environment, and their potential for cleanup or disposal through the use of sorbents. Physical property data on specific gravity, aqueous solubility, liquid surface tension, and viscosity were obtained (see Section 8). The acquisition of these data involved the search of technical handbooks, manuals, reports, and journals. A listing of the literature sources consulted for liquid physical property data is as follows:

Aldrich Catalog/Handbook of Fine Chemicals. Aldrich Chemical Company, Milwaukee, Wisconsin, 1982.

Ambient Water Quality Criteria Documents. EPA 440/5-80, U.S. Environmental Protection Agency, Washington, D.C., 1980.

Bauer, W.H., Borton, D.N., and J.J. Bueloff. Agents, Methods and Devices for Ameliorization of Discharges of Hazardous Chemicals on Water. CG-D038-76, U.S. Coast Guard, 1975.

Chemical Hazards Response Information System (CHRIS). United States Coast Guard, 1978.

Clark, R.S., ed., Tables of Physical Properties. Union Carbide Chemicals Company, South Charleston, West Virginia, 1960.

Fowkes, F.M., ed. Contact Angle, Wettability, and Adhesion. Advances in Chemistry Series 43. American Chemical Society, Washington, D.C., 1964.

Hazardous Chemicals Data, NFPA-49. National Fire Protection Association, 1975.

Herrick, E.C., Carstea, D., and G. Goldgraben. Sorbent Materials for Cleanup of Hazardous Spills. EPA-600/2-82-030, U.S. Environmental Protection Agency, Washington, D.C., 1982.

Ketchen, E.E., and W.E. Porter. Materials Safety Data Sheets the Basis for Control of Toxic Chemicals. ORNL/TM-6981. Oak Ridge National Laboratory, Oak Ridge, Tennessee, 1979.

Lange, N., ed. Lange's Handbook of Chemistry, Ninth edition. McGraw-Hill Book Company, Inc., New York, New York, 1956.

Lawless, E.W., Ferguson, T.L., and A.F. Meiners. Guidelines for the Disposal of Small Quantities of Unused Pesticides. EPA-670/2-75-057, U.S. Environmental Protection Agency, Cincinnati, Ohio, 1975.

MacKison, F.W., Stricoff, R.S. and L.J. Partridge, Jr., eds. NIOSH/OSHA Pocket Guide to Chemical Hazards. USDHHS, USDL, 1980.

Melvold, R.W., et.al. Flammable Hazardous Substances Emergency Response Handbook: Control and Safety Procedures. U.S. Environmental Protection Agency, Cincinnati, Ohio, 1982.

Oil and Hazardous Materials Technical Assistance Data System (OHM/TADS). On-line data base available through the Chemical Information System (CIS) jointly operated by EPA and NIH.

Robertson, L.A., Fuigar, M.F., and L.B. Solsberg. Selection Criteria and Laboratory Evaluation of Oil Spill Sorbents. EPS-4-EC-76-5, Environment Canada, 1976.

Robertson, L.A. Selection Criteria and Laboratory Evaluation of Oil Spill Sorbents: An Update. EPS-4-EP-78-8, Environment Canada, 1978.

Robertson, L.A. Selection Criteria and Laboratory Evaluation of Oil Spill Sorbents: Update II. ES-4-EP-83-4, Environment Canada, 1983.

Rose, A., and E. Rose. The Condensed Chemical Dictionary. Reinhold Publishing Company, New York, New York, 1961.

Sax, I.N. Dangerous Properties of Industrial Materials, Fifth edition. Van Norstrand Reinhold Company, New York, New York, 1979.

Schatzberg, P. Investigation of Sorbents for Removing Oil Spills from Waters. 724110.1/2/1, U.S. Coast Guard, 1971.

Timmermans, J. Physico-Chemical Constants of Pure Organic Compounds. Elsevier Publishing Company, Inc., New York, New York, 1950.

Verschueren, K. Handbook of Environmental Data on Organic Chemicals. Van Nostrand Reinhold Company, New York, New York, 1977.

Ward, C.H., and W.W. Waggett. Research on Hazardous Substances in Support of Spill Prevention Regulations. Department of Environmental Science and Engineering, Rice University, Houston, Texas, 1982.

Weast, R.C., ed. CRC Handbook of Chemistry and Physics, 53 edition. The Chemical Rubber Company, Cleveland, Ohio, 1978.

Windholz, M., ed. The Merck Index, Tenth edition. Merck and Co., Inc.,

Rahway, New Jersey, 1983.

To fill data gaps from the literature sources, data fact sheets and physical property information were solicited from chemical manufacturers of CERCLA liquids. A number of these requests for standard product information were productive in providing previously unobtainable data:

Allied Chemical, P.O. Box 1087R, Nichols 5, Morristown, NJ 07960
American Cyanamid Company, Organic Chemicals Div., One Cyanamid Plaza, Wayne, NJ 07470
Arco Chemical Company, 1500 Market Street, Philadelphia, PA 19101
Ashland Chemical Co., P.O. Box 2219, Columbus, OH 43216
Chemcentral Corporation, 7050 W. 71st Street, Chicago, IL 60638
Columbia Organic Chemicals, 912 Drake Street, Columbia, SC 29205
Denka Chemical Corporation, 8701 Park Place, Houston, TX 77017
Du Pont deNemours & Co., Technical Services Laboratory, Chestnut Run, Wilmington, DE 19898
Eastman Kodak Co., 1669 Lake Avenue, Rochester, NY 14650
Ethyl Corporation, 8000 GSRI Avenue, Baton Rouge, LA 70808
Jim Walter Resources, Inc., P.O. Box 5327, Burmingham, AL 35207
Monsanto Corporation, 800 N. Lindbergh Blvd., St. Louis, MO 63166
Occidental Chemical Corporation, Hooker Chemical Center, Box 728 Niagara Falls, NY 14302
Olin Chemicals, 120 Long Ridge Road, Stanford, CN 06904
Pennwalt Corp., Three Parkway, Philadelphia, PA 19102
Rohm and Haas Company, Independence Mall West, Philadelphia, PA 19105
The Quaker Oats Company, Chemicals Division, P.O. Box 3514, Merchandise Mart Station, Chicago, IL 60654

A literature review was also conducted to obtain relevant sorbent data and to identify commercially available sorbents and their manufacturers. This effort included consideration of the types of sorbent/hazardous liquid properties that would be incorporated into the data sheets. The following publications served as sources of sorbent data:

Bauer, W.H., Borton, D.N., and J.J. Bueloff. Agents, Methods and Devices for Amelioration of Discharges of Hazardous Chemicals on Water. CG-D038-76, U.S. Coast Guard, 1975.

Herrick, E.C., Carstea, D.,and G. Goldgraben. Sorbent Materials for Cleanup of Hazardous Spills., EPA/600/2-82-030, U.S. Environmental Protection Agency, Washington, D.C., 1982.

Robertson, L.A., Fuigar, M.F., and L.B. Solsberg. Selection Criteria and Laboratory Evaluation of Oil Spill Sorbents. EPS-4-EC-76-5, Environment Canada, 1976.

Robertson, L.A. Selection Criteria and Laboratory Evaluation of Oil Spill

Sorbents: An Update. EPS-4-EP-78-8, Environment Canada, 1978.

Robertson, L.A. Selection Criteria and Laboratory Evaluation of Oil Spill Sorbents: Update II. ES-4-EP-83-4, Environment Canada, 1983.

Schatzberg, P. Investigation of Sorbents for Removing Oil Spills from Waters. 724110.1/2/1, U.S. Coast Guard, 1971.

Based on review of the available literature, data required for the project were requested from sorbent manufacturers/producers. Sorbent information, such as sorbent properties, sorbent type, sorbent form, shelf life, applicability to cited spill scenarios, uptake selectivity, application/collection methods, and procurement information, was sought. Application/collection information was also gathered to facilitate the development of Section 3 of the Handbook. In addition, sorbent/hazardous liquid pair data, such as sorption capacity (dry and presaturated), retention capacity, methods for analysis, sorbent compaction and volume increase data, performance parameters (environmental effects and preconditioning requirements), and safety procedures/precautions were requested. Responses generally provided very little new technical data. The following sorbent manufacturers/producers provided information considered useful to the project:

American Vermiculite Corp., 52 Executive Park South, Atlanta, GA 30329
Andersite of California, Inc., 1260 South Goodrich Blvd., Los Angeles, CA 90022
Anderson Cob Mills, Inc., P.O. Box 119, Maumee, OH 43537
Anschutz Minerals Corp., P.O. Box 230, Ochlocknee, GA 31773
Breg-O-S, Inc., Suite 28A, 45 East 89th Street, New York, NY 10028
Clark-Cutler-McDermott Co., P.O. Box 269, Franklin, MA 02038
Clean Water, Inc., P.O. Box 1002, Court House Square, Toms River, NJ 08753
Conwed Corp., 444 Cedar Street, P.O. Box 43237, St. Paul, MN 55164
Crown-Zellerbach, 3720 Grant St., Washougal, WA 98671
Diamond Shamrock Corp., Soda Products Div., 351 Phelps Court, Irving, TX 75061
Dow Chemical Co., 2020 Dow Center, Midland, MI 48640
Eastern Wood Fibers, Inc., 8245 Dorsey Run Road, Jessup, MD 20794
Emco, Inc., 1015 Louisiana Street, Little Rock, AR 72202
Envirosorb Co., 8128 W. 187th S.W., Edmonds, WA 98020
Excel Mineral, 111 S. La Patera Lane, Goleta, CA 93116
Filter Media, Inc., P.O. Box 19156, Houston, TX 77024
Floridin Co., Three Penn Center, Pittsburgh, PA 15235
Georgia Kaolin Co., 433 N. Broad Street, Elizabeth, NJ 07207
Geotech Development Corp., 630 Valley Forge Plaza Bldg., King of Prussia, PA 19406
J.V. Manufacturing, ARG-Dri Div., P.O. Box 371, De Pere, WI 54115
Lowe's, Inc., 21525 Allegheny Street, Cassopolis, MI 49031
Oil-Dri Corporation, 520 N. Michigan Avenue, Chicago, IL 60611
Petro-Trap, P.O. Box 157, Westport, MA 02790
R.B.H. Cybernetics Ltd., P.O. Box 4205, Postal Station "A",

Victoria, B.C., Canada V8X 3X8
Radecca, Inc., 8501 Mo-Pac Blvd., P.O. Box 9948, Austin, TX 78766
Sea Clean Inc., 7000 S.W. 62nd Avenue, Suite 555, Miami, FL 33143
Sorbent Products Co., P.O. Box 174, Maplewood, NJ 07040
Spill Control Co., Sorbent Sciences Div., 828 N. Grand Avenue, Covina, CA 91724
Terra Nova Power & Development Ltd., P.O. Box 13531, St. Johns, Newfoundland, Canada A1B 4B8
3M, Occupational Health and Safety Products Div., 3M Center, St. Paul, MN 55144
Waverly Minerals, 555 City Line Avenue, Bala Cynwyd, PA 19004

8. CERCLA Liquid Chemical Information

This section presents physical property data for the CERCLA liquids. Four physical properties, namely, specific gravity, solubility in water, viscosity, and surface tension are considered significant in influencing the sorption of a liquid by a sorbent. These data were used in estimating sorption capacity values based on extrapolations from experimental data (see Section 2) and are presented in tabular form in Table 8-1 so that the user can make comparisons among CERCLA liquids with similar physical properties for the same purpose. Additionally, the data on liquid specific gravity when used in conjunction with sorbent density (from the Sorbent Data Sheets) and sorption capacity (from Table 2-1) can be used to calculate the volume of sorbent/hazardous liquid after cleanup for the purpose of estimating disposal needs.

Table 8-1 is organized alphabetically by CERCLA liquid chemical class followed by the hazardous liquid CERCLA name, alphabetically in each class, and the CAS Registry number. The remaining columns contain the physical property data. The qualitative solubility data is either "soluble" (i.e., solubility equal to or greater than 1%) or "insoluble" (i.e., solubility less than 1%). Each of the four salient physical properties, their significance and the manner in which each influences cleanup by sorption are discussed in the following paragraphs.

SPECIFIC GRAVITY

The specific gravity of a substance is the ratio of its density to that of water and, as a result, is unitless. Since density varies with temperature, the temperatures of the substance and the water are sometimes given. Specific gravity is typically expressed in terms of water at 4 C, the temperature at which water has its greatest density.

A specific gravity greater than unity indicates that the substance is denser than water and will sink in water. A substance with a specific gravity less than unity will float on water. The actual distribution in the water column of a substance whose specific gravity is near unity will depend on the influence of temperature and the density of the water, as well as currents and wave action. For example, a spill of a substance with a specific gravity of 0.999 would actually sink in water warmer than about 16 C, albeit quite slowly. On the other hand, a spill of a substance with a specific gravity of 1.02 would be expected to sink in fresh water but might actually float in sea water. The specific gravities of the CERCLA liquids range from 0.68 to 13.5.

Specific gravity is not considered to influence the rate of sorption as much as surface tension and viscosity. However, if two liquids exhibit similar surface tensions and viscosities, their sorption capacities will be influenced by specific gravity.

SOLUBILITY IN WATER

The solubility of a substance in water is a measure of the maximum quantity of the substance that can dissolve in a given amount of water under a particular set of conditions. Water temperature and the presence of other solutes influence solubility. Aqueous solubility generally increases for liquids as water temperature increases. The presence of other solutes depresses solubility. For the purposes of this Manual, solubility is expressed in parts per million on a weight basis. The solubilities of the CERCLA liquids range from 0.02 ppm to complete miscibility.

The solubility of a substance will influence its behavior when spilled into water. In addition, the relative sizes of the spill and the water body into which the spill occurs will also influence spill behavior. For example, a small spill of a substance with a low aqueous solubility into a large body of water may dissolve totally while a spill of a substance of a much higher solubility into a small body of water may produce two phases.

The degree of turbulence and other mixing processes in the body of water will also influence the rate of dissolution. Vigorous mixing will enhance the rate of dissolution, while quiescent conditions will protract the time required for equilibrium conditions to be achieved.

The aqueous solubility of a substance may be used as an indicator of its degree of polarity which may influence the choice of sorbent. Substances that are highly soluble in water have polar functional groups and they generally would be easily sorbed by hydrophilic sorbents. Of course, in an aqueous system a hydrophilic sorbent would also sorb water. On the other hand, a substance of very low solubility in water is nonpolar and would best be sorbed by oleophilic or hydrophobic sorbents.

SURFACE TENSION

Surface tension (or surface free energy) is an internal force that causes the surface of a liquid to curve near the walls of its container and to form rounded drops when not in a container. Surface tension is the force per unit length on the plane of the surface, usually expressed in dynes/cm. The range of surface tensions of the CERCLA liquids is 5 to 100 dynes/cm, excluding mercury (484.0 dynes/cm). For reference, the surface tension of water is 72 dynes/cm at 25 C.

Surface tension determines the degree to which a liquid substance spreads on or "wets" the surface of a solid. For the same solid surface, a liquid of high surface tension will not spread as much as a liquid with low surface tension. For example, mercury, with a surface tension of 484 dynes/cm at 15 C does not spread on a clean glass surface, but maintains a droplet shape. On the other hand, water easily spreads into a film on a clean glass surface.

Another example is the nonwetting of wax paper by water compared to a liquid with a smaller surface tension, like ethanol (surface tension = 22.3 dynes/cm).

Significant wetting of a liquid on a solid surface will only occur when the surface energy of the solid is greater than the surface tension of the liquid. For example, Teflon has a very low surface energy and few liquids can wet or spread on such a surface.

Surface tension, in part, determines whether or not a particular sorbent will be effective. In order for the sorbent to be effective the hazardous liquid must wet the sorbent surface. If the liquid cannot wet the sorbent, the sorbent is rendered ineffective because the liquid blocks the sorbent pores, preventing liquid from being drawn into the interstices of the sorbent.

Sorbents constructed of materials with surface free energies lower than the surface tension of water (e.g., polyethylene) will selectively sorb liquids of lower surface tension in a two-phase liquid system where one of the phases is water. Sorbents with surface free energies greater than the surface tension of water can be wetted with water, and if this happens prior to contact with a spilled material of lower surface tension relative to water, the sorbent will be rendered ineffective because of its saturation with water.

VISCOSITY

The viscosity of a liquid is a measure of the forces that work against movement or flow when a shearing stress is applied. A typical unit of viscosity is the centipoise (cp). The range of viscosities of the CERCLA liquids is 0.01 to 58 cp. For reference purposes, the viscosity of water is 1 cp at 20°C, and the viscosity of 10-weight motor oil is about 100 cp at 20°C. Viscosity is inversely related to temperature. For example, water is half as viscous at 55°C as it is at 20°C. The relative influence of temperature on viscosity is greater for highly viscous substances than for less viscous substances.

Viscosity is an important consideration during the cleanup of a spill with sorbents. While viscosity plays a role in the maximum sorption capacity achievable by a sorbent, its most significant effect is on the rate of sorption (Ref. 9). In other words, it will take longer to sorb a substance at 0°C, than it will at 20°C, but the amount of hazardous liquid per unit of sorbent will be nearly the same (note that an increase in temperature also decreases surface tension which may influence sorption capacity).

Substances with relatively low viscosities will have a greater tendency to drain from the sorbent during collection compared with more viscous substances. The degree to which this happens depends also on the properties of the sorbent, such as pore size and surface area.

TABLE 8-1. CERCLA HAZARDOUS LIQUID PHYSICAL PROPERTIES

Guide Number	CERCLA Class	Hazardous Liquid Name	CAS No.	Specific Gravity	Solubility (ppm)	Surface Tension (dynes/cm)	Viscosity (cp)
1	Acidic compounds, inorganic	Hydrochloric acid	7647-01-0	1.19	832000	70.0	25.4
1	Acidic compounds, inorganic	Hydrocyanic acid	74-90-8	0.69	Misc	18.2	0.29
1	Acidic compounds, inorganic	Hydrofluoric acid	7654-39-3	0.99	Misc	8.6	0.3
1	Acidic compounds, inorganic	Nitric acid	7697-37-2	1.50	Misc	42.7	1.7
1	Acidic compounds, inorganic	Nitrogen dioxide	10102-44-0	1.45	Dec	27.5	
1	Acidic compounds, inorganic	Phosphoric acid	7664-38-2	1.87	Misc		28.5
1	Acidic compounds, inorganic	Selenious acid	7783-00-8	3.00	270000		
1	Acidic compounds, inorganic	Sulfuric acid	7664-93-9	1.84	Misc	55.1	25.4
2	Acidic compounds, organic	Acetic acid	64-19-7	1.05	Misc	27.8	1.3
2	Acidic compounds, organic	Acetic anhydride	108-24-7	1.08	120000	32.7	0.9
2	Acidic compounds, organic	Acrylic acid	79-10-7	1.10	Misc	28.1	1.2
2	Acidic compounds, organic	Butyric acid	107-92-6	0.96	Misc	26.7	1.5
2	Acidic compounds, organic	Chlorosulfonic acid	7790-94-5	1.76	Dec		
2	Acidic compounds, organic	Dodecylbenzensulfonic acid	27176-87-0	1.02	Misc		
2	Acidic compounds, organic	Formic acid	64-18-6	1.22	Misc	37.6	1.8
2	Acidic compounds, organic	Propionic acid	79-09-4	0.99	Misc	27.2	1.1

(continued)

TABLE 8-1. (continued)

Guide Number	CERCLA Class	Hazardous Liquid Name	CAS No.	Specific Gravity	Solubility (ppm)	Surface Tension (dynes/cm)	Viscosity (cp)
2	Acidic compounds, organic	Propionic anhydride	123-62-6	1.01	Dec	30.0	1.1
3	Alcohols & glycols	1-Butanol	71-63-3	0.81	77000	24.6	2.9
3	Alcohols & glycols	Allyl alcohol	107-18-6	0.82	Misc	25.8	1.4
3	Alcohols & glycols	Isobutyl alcohol	78-83-1	0.81	40000	23.0	4.7
3	Alcohols & glycols	Methanol	67-56-1	0.79	Misc	22.6	0.6
3	Alcohols & glycols	Propargyl alcohol	107-19-7	0.97	Misc		1.7
3	Alcohols & glycols	Xylenol	1300-71-6	1.01	Sol	30.0	
4	Aldehydes	Acetaldehyde	75-07-0	0.78	Misc	21.2	0.2
4	Aldehydes	Acrolein	107-02-8	0.84	400000	24.0	0.3
4	Aldehydes	Chloral	75-87-6	1.51	14740	25.3	
4	Aldehydes	Chloroacetaldehyde	107-20-0	1.19	10000		
4	Aldehydes	Crotonaldehyde	4170-30-3	0.85	153000		
4	Aldehydes	Endrin aldehyde	930-55-2	>1	Insol		
4	Aldehydes	Furfural	98-01-1	1.16	83000	43.5	1.5
4	Aldehydes	Glycidyladehyde	765-34-4	1.10	Sol		
4	Aldehydes	Paraldehyde	123-63-7	0.99	120000	25.9	1.0
5	Aliphatic hydrocarbons	Cyclohexane	110-82-7	0.78	55	25.5	1.0
5	Aliphatic hydrocarbons	1,3-Pentadiene	504-60-9	0.69	Insol		
5	Aliphatic hydrocarbons	Isoprene	78-79-5	0.68	100	16.9	0.2
6	Aliphatics, halogenated	1,1,1,2-Tetrachloroethane	630-20-6	1.53	2600		
6	Aliphatics, halogenated	1,1,2,2-Tetrachloroethane	79-34-5	1.59	2857	36.0	1.8
6	Aliphatics, halogenated	1,1,2-Trichloroethane	79-00-5	1.44	4500	22.0	1.2

(continued)

TABLE 8-1. (continued)

Guide Number	CERCLA Class	Hazardous Liquid Name	CAS No.	Specific Gravity	Solubility (ppm)	Surface Tension (dynes/cm)	Viscosity (cp)
6	Aliphatics, halogenated	1,1-Dichloroethane	75-34-3	1.17	5500	24.8	0.5
6	Aliphatics, halogenated	1,2-Dibromo-3-chloropropane	96-12-8	2.09	Insol		
6	Aliphatics, halogenated	1,2-Trans-dichloroethylene	156-60-5	1.27	600		
6	Aliphatics, halogenated	1,3-Dichloropropene	542-75-6	1.22	2800	31.2	0.8
6	Aliphatics, halogenated	1.4-Dichloro-2-butene	764-41-0	1.19	2000	24.0	1.0
6	Aliphatics, halogenated	2,2-Dichloropropionic acid	75-99-0	1.39	450000		
6	Aliphatics, halogenated	Acetyl bromide	506-96-7	1.52	Dec		
6	Aliphatics, halogenated	Acetyl chloride	75-36-5	1.10	Dec	26.7	
6	Aliphatics, halogenated	Allyl chloride	107-05-1	0.94	100	28.9	0.4
6	Aliphatics, halogenated	Benzenesulfonyl chloride	98-09-9	1.38	Insol		
6	Aliphatics, halogenated	Bis(2-chloroethoxy)methane	111-91-1	1.23	81000		3.9
6	Aliphatics, halogenated	Bromoform	75-25-2	2.89	3190	41.5	1.9
6	Aliphatics, halogenated	Carbon tetrachloride	56-23-5	1.59	800	27.0	1.0
6	Aliphatics, halogenated	Chlorodibromomethane	124-48-1	2.38	Insol		
6	Aliphatics, halogenated	Chloroethane	75-00-3	0.92	5700	19.5	0.8
6	Aliphatics, halogenated	Chloroform	67-66-3	1.48	7365	27.1	0.6

(continued)

TABLE 8-1. (continued)

Guide Number	CERCLA Class	Hazardous Liquid Name	CAS No.	Specific Gravity	Solubility (ppm)	Surface Tension (dynes/cm)	Viscosity (cp)
6	Aliphatics, halogenated	Cyanogen chloride	506-77-4	1.19	30000	24.6	
6	Aliphatics, halogenated	Dichlorobromomethane	75-27-4	1.97	4700		
6	Aliphatics, halogenated	Dichloropropane (all isomers)	26638-19-7	1.16	2700	29.0	0.9
6	Aliphatics, halogenated	Dichloropropene (all isomers)	26952-23-8	1.22	Insol	31.2	
6	Aliphatics, halogenated	Dichloropropene-dichloropropane mixture	8003-19-8	1.19	1000	28.9	
6	Aliphatics, halogenated	Dimethyl carbamoyl chloride	79-44-7	1.68	Sol		
6	Aliphatics, halogenated	Ethylene dibromide	106-93-4	2.17	4310	38.8	1.7
6	Aliphatics, halogenated	Ethylene dichloride	107-06-2	1.26	8690	32.2	0.8
6	Aliphatics, halogenated	Hexachlorobutadiene	87-68-3	1.67	5	35.3	2.5
6	Aliphatics, halogenated	Hexachlorocyclopentadiene	77-47-4	1.71	0.8	37.5	
6	Aliphatics, halogenated	Hexachloropropene	1888-71-7	1.76	Insol	38.1	
6	Aliphatics, halogenated	Methyl chloroform	71-55-6	1.34	4400	25.4	
6	Aliphatics, halogenated	Methyl iodide	74-88-4	2.28	14000	25.8	0.5
6	Aliphatics, halogenated	Methylene bromide	74-95-3	2.50	11700		1.1
6	Aliphatics, halogenated	Methylene chloride	75-09-2	1.32	20000	26.5	0.5
6	Aliphatics, halogenated	Pentachloroethane	76-01-7	1.67	500		

(continued)

TABLE 8-1. (continued)

Guide Number	CERCLA Class	Hazardous Liquid Name	CAS No.	Specific Gravity	Solubility (ppm)	Surface Tension (dynes/cm)	Viscosity (cp)
6	Aliphatics, halogenated	Propylene dichloride	78-87-5	1.16	2700	29.0	0.9
6	Aliphatics, halogenated	Tetrachloroethylene	127-18-4	1.63	150	31.3	0.9
6	Aliphatics, halogenated	Trichloroethylene	79-01-6	1.46	1000	29.3	0.6
6	Aliphatics, halogenated	Trichloromonofluromethane	75-69-4	1.49	1100	23.0	0.4
6	Aliphatics, halogenated	Vinylidene chloride	75-35-4	1.21	400	24.0	0.5
7	Amides, anilides, & imides	Octamethylpyprophosphoramide	152-16-9	1.09	Misc		
8	Amines, alkyl	1,2-Propylenimine	75-55-8	0.80	Misc	25.0	
8	Amines, alkyl	Butylamine	109-73-9	0.74	Misc	19.7	0.5
8	Amines, alkyl	Diethylamine	109-89-7	0.71	Misc	16.4	0.4
8	Amines, alkyl	Dipropylamine	142-84-7	0.74	Misc	6.6	0.5
8	Amines, alkyl	Ethylenediamine	107-15-3	0.90	Misc	40.8	1.7
8	Amines, alkyl	Ethyleneimine	151-56-4	0.83	Misc	34.5	
8	Amines, alkyl	Monoethylamine	75-04-7	0.69	Misc	20.4	0.3
8	Amines, alkyl	N-Propylamine	107-10-8	0.72	Misc	57.7	0.4
8	Amines, alkyl	Triethylamine	121-44-8	0.73	15000	20.7	0.9
9	Amines, aryl	Aniline	62-53-3	1.02	35000	45.5	4.4
9	Amines, aryl	2,4,5-T Amines	2008-46-0	1.75	Insol		
9	Amines, aryl	2-Picoline	109-06-8	0.95	Misc		
9	Amines, aryl	alpha,alpha-Dimethylphenethylamine	122-09-8	<1	Sol		
9	Amines, aryl	Methapyriline	91-80-5	>1	Sol		
9	Amines, aryl	Nicotine & Salts	54-11-5	1.01	Misc	38.6	
9	Amines, aryl	Pyridine	110-86-1	0.98	Misc	38.0	1.0

(continued)

TABLE 8-1. (continued)

Guide Number	CERCLA Class	Hazardous Liquid Name	CAS No.	Specific Gravity	Solubility (ppm)	Surface Tension (dynes/cm)	Viscosity (cp)
9	Amines, aryl	Quinoline	91-22-5	1.10	60000	45.0	0.6
10	Aromatic hydrocarbons	Benzene	71-43-2	0.88	1780	28.9	0.7
10	Aromatic hydrocarbons	Creosote	8001-58-9	1.08	7122	15.0	12.0
10	Aromatic hydrocarbons	Cumene	98-82-8	0.86	50	28.2	0.7
10	Aromatic hydrocarbons	Dihydrosafrole	94-58-6	1.07	Insol		
10	Aromatic hydrocarbons	Ethylbenzene	100-41-4	0.87	152	29.2	0.7
10	Aromatic hydrocarbons	Isosafrole	120-58-1	1.12	Insol		4.0
10	Aromatic hydrocarbons	Safrole	94-59-7	1.10	Insol		
10	Aromatic hydrocarbons	Styrene	100-42-5	0.91	280	32.2	0.8
10	Aromatic hydrocarbons	Toluene	108-88-3	0.87	515	29.0	0.6
10	Aromatic hydrocarbons	Xylene	1330-20-7	0.87	186	28.5	0.7
11	Aromatics, halogenated	1,2,4-Trichlorobenzene	120-82-1	1.45	Insol		
11	Aromatics, halogenated	2-Chlorophenol	95-57-8	1.26	28500	42.3	4.1
11	Aromatics, halogenated	Benzal chloride	98-87-3	1.25	Insol		
11	Aromatics, halogenated	Benzotrichloride	98-07-7	1.38	Dec		2.6
11	Aromatics, halogenated	Benzoyl chloride	98-88-4	1.21	Dec	35.3	1.2
11	Aromatics, halogenated	Benzyl chloride	100-44-7	1.10	33	37.6	0.01
11	Aromatics, halogenated	Chlorobenzene	108-90-7	1.11	500	33.6	0.8
11	Aromatics, halogenated	Dichlorobenzene (all isomers)	25321-22-6	>1	Insol		
11	Aromatics, halogenated	m-Dichlorobenzene	541-73-1	1.29	123		
11	Aromatics, halogenated	o-Dichlorobenzene	95-50-1	1.30	100	37	1.3
11	Aromatics, halogenated	Polychlorinated biphenyls	1336-36-3	1.30	0.2	45.3	10.0
12	Basic compounds	Ammonium hydroxide	1336-21-6	0.89	Misc		
12	Basic compounds	Sodium hypochlorite (in aqueous solution)	10022-70-5	1.06	293000		

(continued)

TABLE 8-1. (continued)

Guide Number	CERCLA Class	Hazardous Liquid Name	CAS No.	Specific Gravity	Solubility (ppm)	Surface Tension (dynes/cm)	Viscosity (cp)
13	Cyanates & isocyanates	Methyl isocyanate	624-83-9	<1	Sol		
13	Cyanates & isocyanates	Toluene diisocyanate	584-84-9	1.22	Dec	25.0	1.8
14	Cyanides & nitriles	3-Chloropropionitrile	542-76-7	1.14	43062		
14	Cyanides & nitriles	Acetone cyanohydrin	75-86-5	0.93	Misc	30.4	5.0
14	Cyanides & nitriles	Acetonitrile	75-05-8	0.79	Misc	29.3	0.3
14	Cyanides & nitriles	Acrylonitrile	107-13-1	0.81	73500	27.3	0.3
14	Cyanides & nitriles	Benzonitrile	100-47-0	1.01	10000	34.7	1.2
14	Cyanides & nitriles	Ethyl cyanide	107-12-0	0.78	119000	27.3	0.4
14	Cyanides & nitriles	Methacrylonitrile	126-98-7	0.80	16200	24.4	0.4
15	Esters & ethers	2,4,5-TP Acid esters	NA	>1	Insol		
15	Esters & ethers	2,4,5-T Esters	93-79-8	1.25	10		
15	Esters & ethers	2,4-D Esters	94-11-1	1.25	10		
15	Esters & ethers	1,4-Dioxane	123-91-1	1.03	Misc	33.4	1.2
15	Esters & ethers	Amyl acetate	628-63-7	0.88	1800	12.0	1.0
15	Esters & ethers	Bis(2-ethylhexyl) phthalate	117-81-7	0.98	2000	15.0	57.0
15	Esters & ethers	Butyl acetate	123-86-4	0.88	14000	24.6	0.7
15	Esters & ethers	Butyl benzyl phthalate	85-68-7	1.12	3		46.4
15	Esters & ethers	Di-n-butylphthalate	84-74-2	1.04	400	34.0	20.0
15	Esters & ethers	Di-n-octylphthalate	117-84-0	0.99	3	15.0	58.0
15	Esters & ethers	Diethyl phthalate	84-66-2	1.23	896	37.5	12.0
15	Esters & ethers	Dimethyl phthalate	131-11-3	1.19	4300		17.2
15	Esters & ethers	Ethyl acetate	141-78-6	0.90	90000	23.9	0.5
15	Esters & ethers	Ethyl acrylate	140-88-5	0.94	20000	25.0	0.6
15	Esters & ethers	Ethyl ether	60-29-7	0.71	84300	17.0	0.2
15	Esters & ethers	Ethyl methacrylate	97-63-2	0.91	Insol		

(continued)

TABLE 8-1. (continued)

Guide Number	CERCLA Class	Hazardous Liquid Name	CAS No.	Specific Gravity	Solubility (ppm)	Surface Tension (dynes/cm)	Viscosity (cp)
15	Esters & ethers	Ethyl methanesulfonate	62-50-0	1.20	Sol		
15	Esters & ethers	Ethyl-4,4'dichlorobenzilate	510-15-6	>1	Insol		
15	Esters & ethers	Furan	110-00-9	0.94	10000	24.1	0.4
15	Esters & ethers	Methyl chlorocarbonate	79-22-1	1.22	Sol	26.0	
15	Esters & ethers	Methyl methacrylate	80-62-6	0.95	15000	28.0	
15	Esters & ethers	Tetrahydrofuran	109-99-9	0.89	Misc	28.0	0.5
15	Esters & ethers	Vinyl acetate	108-05-4	0.93	25000	24.0	0.5
16	Ethers, halogenated	2-chloroethyl vinyl ether	110-75-8	1.05	15000	28.8	0.8
16	Ethers, halogenated	4-Bromophenyl phenyl ether	101-55-3	1.42	5		
16	Ethers, halogenated	4-Chlorophenyl phenyl ether	7005-72-3	>1	3		
16	Ethers, halogenated	Bis(2-chloroethyl) ether	111-44-4	1.22	10200	37.9	3.0
16	Ethers, halogenated	Bis(2-chloroisopropyl) ether	108-60-1	1.11	1700		2.3
16	Ethers, halogenated	Bis(chloromethyl) ether	542-88-1	1.31	22000		
16	Ethers, halogenated	Chloromethyl methyl ether	107-30-2	1.07	5000	30.0	
16	Ethers, halogenated	Epichlorohydrin	106-89-8	1.18	66	37.0	1.1
17	Halides, inorganic	Antimony pentachloride	7647-18-9	2.35	Dec	15.0	
17	Halides, inorganic	Arsenic trichloride	7784-34-1	2.16	Dec	20.0	1.1
17	Halides, inorganic	Phosphorus oxychloride	10025-87-3	1.68	Dec		1.0
17	Halides, inorganic	Phosphorus trichloride	7719-12-2	1.57	Dec	25.6	0.5
17	Halides, inorganic	Sulfur monochloride	12771-08-3	1.68	Dec		1.2
18	Heavy Metals	Diethylarsine	692-42-2	1.13	Insol		
18	Heavy Metals	Lead fluoborate (51% aqueous solution)	13814-96-5	>1	Misc		
18	Heavy Metals	Mercury	7439-97-6	13.5	0.02	484.0	1.6
18	Heavy Metals	Nickel carbonyl	13463-39-3	1.32	180	15.9	

(continued)

TABLE 8-1. (continued)

Guide Number	CERCLA Class	Hazardous Liquid Name	CAS No.	Specific Gravity	Solubility (ppm)	Surface Tension (dynes/cm)	Viscosity (cp)
18	Heavy Metals	Tetraethyl lead	78-00-2	1.63	30	28.5	
19	Hydrazines & hydrazides	1,1-Dimethylhydrazine	57-14-7	0.79	Misc	28.0	
19	Hydrazines & hydrazides	1,2-Dimethylhydrazine	540-73-8	0.83	Misc		
19	Hydrazines & hydrazides	Hydrazine	302-01-2	1.00	Misc	91.5	1.0
19	Hydrazines & hydrazides	Methyl hydrazine	60-34-4	0.88	Misc	34.3	0.9
19	Hydrazines & hydrazides	N,N'-Diethylhydrazine	1615-80-1	0.80	Sol		
20	Ketones	Acetone	67-64-1	0.79	Misc	23.7	0.3
20	Ketones	Acetophenone	98-85-2	1.03	5500	39.8	1.6
20	Ketones	Bromoacetone	598-31-2	1.63	Sol		
20	Ketones	Cyclohexanone	108-94-1	0.95	150000	34.0	2.2
20	Ketones	Isophorone	78-59-1	0.92	12000	32.3	2.6
20	Ketones	Methyl ethyl ketone	78-93-3	0.81	Misc	24.6	0.4
20	Ketones	Methyl isobutyl ketone	108-10-1	0.80	19100	23.6	0.5
21	Nitro/nitroso compounds	2-Nitropropane	79-46-9	0.98	16000	30.0	
21	Nitro/nitroso compounds	Nitrobenzene	98-95-3	1.21	1900	43.9	2.0
21	Nitro/nitroso compounds	Tetranitromethane	509-14-8	1.65	Insol		1.8
21	Nitro/nitroso compounds	N-Nitroso-n-methylurea	684-93-5	>1	Sol		
21	Nitro/nitroso compounds	N-Nitrosodi-n-propylamine	621-64-7	0.92	9900		
21	Nitro/nitroso compounds	N-Nitrosodiethanolamine	1116-54-7	1.00	Sol		

(continued)

TABLE 8-1. (continued)

Guide Number	CERCLA Class	Hazardous Liquid Name	CAS No.	Specific Gravity	Solubility (ppm)	Surface Tension (dynes/cm)	Viscosity (cp)
21	Nitro/nitroso compounds	N-Nitrosodiethylamine	55-18-5	0.94	Sol		
21	Nitro/nitroso compounds	N-Nitrosodimethylamine	62-75-9	1.00	Misc		
21	Nitro/nitroso compounds	N-Nitrosomethylvinylamine	4549-40-0	>1	Sol		
21	Nitro/nitroso compounds	N-Nitrosopyrrolidine	930-55-2	1.10	Sol		
22	Organophosphates	Diazinon	333-41-5	1.12	40	35.0	
22	Organophosphates	Dichlorvos	62-73-7	1,41	10000		
22	Organophosphates	Diethyl-p-nitrophenyl phosphate	311-45-5	1.27	12730		
22	Organophosphates	Diisopropyl fluorophosphate	55-91-4	1.05	15400		
22	Organophosphates	Ethion	563-12-2	1.22	2		
22	Organophosphates	Hexaethyl tetraphosphate	757-58-4	1.26	Misc		
22	Organophosphates	Malathion	121-75-5	1.23	145	37.1	
22	Organophosphates	Mevinphos	7786-34-7	1.25	Misc		
22	Organophosphates	O,O-Diethyl-O-pyrazinyl phosphorothioate	297-97-2	1.25	1140		
22	Organophosphates	O,O-Diethyl-S-methyl dithiophosphate	3268-58-2	>1	Insol		
22	Organophosphates	Parathion	56-38-2	1.26	20	39.2	15.3
22	Organophosphates	Tetraethyl pyrophosphate	107-49-3	1.18	Misc		
22	Organophosphates	Tetraethyldithiopyrophosphate	3689-24-5	1.19	25		
23	Oxides, alkylene	Ethylene oxide	75-21-8	0.87	Misc	24.3	0.3
23	Oxides, alkylene	Propylene oxide	75-56-9	0.83	650000	24.5	0.3
23	Oxides, alkylene	2,2'-Bioxirane	1464-53-5	1.11	Dec	39.7	1.5
24	Peroxides	2-Butanone peroxide	1338-23-4	1.17	Sol		14.8
24	Peroxides	alpha,alpha-Dimethylbenzylhydroperoxide	80-15-9	1.03	Insol	25.0	
25	Phenols & cresols	Cresol	1319-77-3	1.03	19600	37.0	20.8

(continued)

TABLE 8-1. (continued)

Guide Number	CERCLA Class	Hazardous Liquid Name	CAS No.	Specific Gravity	Solubility (ppm)	Surface Tension (dynes/cm)	Viscosity (cp)
25	Phenols & cresols	Thiophenol	108-98-5	1.08	Insol		
26	Sulfates & sulfites	Dimethyl sulfate	77-78-1	1.33	28000	40.1	1.7
26	Sulfates & sulfites	Propargite	2312-35-8	1.09	0.5		
27	Sulfides & mercaptans	Ammonium sulfide (in aqueous solution)	12135-76-1	1.10	Misc		
27	Sulfides & mercaptans	Carbon disulfide	75-15-0	1.26	2940	32.3	0.4

Legend:

Dec = Decomposes
Insol = Insoluble (less than 1%)
Misc = Miscible (soluble in any proportion)
Sol = Soluble (greater than or equal to 1%)

9. Sorbent Selection Rationale

INTRODUCTION

The "Sorbent Selection and Use Guides" (Guides) contain sorbent lists that are ranked according to estimated performance for each of several applicable spill scenarios. A sorbent's ranking reflects the relative estimated performance of the sorbent in the particular scenario. The sorbent performance ranking procedure is described in this section, and the user is advised to become familar with the assumptions, estimations, computations, limitations, and constraints that influenced the estimated performance ratings in the Guides.

The sorbent performance ranking procedure consisted of four phases. First, the conceptual framework for making performance comparisons was developed. Next, sorbent performance evaluation factors and criteria were selected for each spill scenario that was included in the Manual. Third, a computaton process was established so that an overall numerical performance rating could be assigned for the sorbent-chemical combinations for each applicable spill scenario. Finally, the existence of an overall numerical performance rating was used to make inter-sorbent performance comparisons for each spill scenario.

It was recognized that cost and availability are sometimes, but not always, important factors in the selection of a sorbent. It appeared that, as a general rule, when the time available to select and use the sorbent decreased, the importance of cost as a selection factor decreased and availability increased. Therefore, it was decided that cost would not be a performance factor, while availability would be, for the Landspill and Floating Spill scenarios, where response time would be critical in many cases. On the other hand, cost was used as a performance factor, and availability was not, for the Landfill scenario, where it was thought that time would not be as restricted and cost considerations would assume greater importance.

The user who is concerned about costs for the Landspill and Floating Spill Scenarios is advised to regard the sorbents listed in each Guide as an unranked list of up to six sorbents that are expected to be acceptable, to varying degrees, for the scenario. The user should then evaluate and compare the overall cost of using each of the sorbents listed by applying guidance and data from the appropriate sections of the Manual.

DEVELOPMENT OF A FRAMEWORK FOR MAKING SORBENT PERFORMANCE COMPARISONS

Prior to comparing the performance of various sorbents, it was necessary to develop a framework for making comparisons. The framework was constructed by:

(1) Identifying release scenarios for which sorbents would be potentially useful and which could be addressed in the Manual.

Three major scenarios (Landspills, Floating Spills, and Landfill) were selected for inclusion in the Manual. The Landspill scenario was subdivided into Small (<500 gallons) and Medium (500 to 5000 gallons) Landspills. The use of sorbents for spill volumes greater than 5000 gallons was thought rather unlikely due to the increased feasibility of recovery by pumping. The Floating Spill scenario was subdivided into Non-Flowing Small Pond, Non-Flowing Large Lake, Flowing Small Stream, and Flowing Large River, since flowrate and size of the waterbody were judged to have significant effects on the sorbent application and collection options available to the user. A Landfill scenario was included that involved using the sorbent to immobilize 1000 gallons of hazardous liquid, transport it, and dispose of it in a permitted landfill. The use of sorbents for chemicals dispersed or dissolved in water was not addressed due to limitations on project resources and the lack (except for activated carbon) of available data.

(2) Identifying hazardous substances for which sorbents would be potentially useful.

CERCLA-designated liquid hazardous substances were chosen as the chemicals having the greatest potential for being controlled by the use of sorbents, and 212 of these liquids were identified early in the project. These chemicals were grouped into 27 classes based on functional groups. The assumption was made, unless there were information or data indicating otherwise, that the chemicals in a class would interact with sorbents in a similar manner.

(3) Identifying sorbents that would be potentially useful for spill control operations.

Twenty-six specific sorbents have been identified in the Manual. These sorbents were placed into 12 generic sorbent classes, based upon their composition. (Some potential sorbents were not included because they did not belong in one of the 12 generic sorbent classes <u>and</u> sorption capacity data, either literature or experimental, for them were not available.) A "representative" sorbent was chosen for each class and it was assumed, unless there were information or data to indicate otherwise, that the sorbents in a class would interact with the same chemical in a similar manner. Subsequent sorbent performance comparisons were based on the performance of the "representative" sorbent with the "representative" chemical.

SELECTION OF SORBENT PERFORMANCE FACTORS FOR EACH SCENARIO

After the Scenario/Chemical/Sorbent framework was established, sorbent performance factors were identified that were regarded as important indicators of sorbent performance for each of the spill scenarios in the Manual. The sorbent performance factors were of two types:

1. Pass/Fail Factors, which were considered mandatory for minimal acceptable performance. These factors had to be satisfied by the sorbent or it was eliminated from further consideration for that spill scenario.

2. Ratable Performance Factors, which were recognized as having an important influence on sorbent performance, but low performance in one particular area would not necessarily eliminate a sorbent from consideration for use. These factors were evaluated on a quantitative or semi-quantitative basis for each scenario. Also, some ratable performance factors are not as important as others in affectng sorbent performance. This difference in importance was accounted for to some degree by sub-dividing the Ratable Performance Factors into two sub-classes:

   a) Primary factors, which generally predominate in determining the sorbent's performance; and,

   b) Secondary factors, which generally are not as significant compared to primary factors, but which can have an important effect on performance under certain conditions and must be considered during the sorbent selection process. Primary factors were given twice the weight of the secondary factors in the sorbent performance evaluation process.

The Sorbent Performance Factors are listed below for each of the seven spill scenarios covered in the Manual.

TABLE 9-1. SORBENT PERFORMANCE FACTORS

Scenarios	Pass/Fail Factors	Ratable Factors	
		Primary	Secondary
1. Landspill a. Small b. Medium	Compatibility Suitability of Form Data	Application Collection Availability	Reusability Incinerability Wind Scatter Resistance Precipitation Effects Environmental Effects Dense Ground Cover Effects Rugged Terrain Effects
2. Floating Spills a. Small, non-flowing b. Small, flowing c. Large, non-flowing d. Large, flowing	Compatibility Suitability of Form Data Specific Gravity	Application Collection Availability Hydrophobicity	Reusability Incinerability Wind Scatter Resistance Precipitation Effects Environmental Effects
3. Landfill	Compatibility Suitability of Form Data	Disposal Cost	

SORBENT PERFORMANCE RANKING PROCEDURE - LANDSPILL SCENARIO

Pass/Fail Evaluation

For the Landspill Scenario, sorbents were eliminated from further consideration for a particular chemical class if they were incompatible or not wet by the representative chemical. Sorbent booms were also not considered for this scenario. A sorbent-chemical combination could also be eliminated if insufficient data were available to develop performance ratings.

Primary Factor Evaluation

The narrative below describes how the Application, Collection, and Availability ratings were derived for the Landspill Scenario. Tables 9-2 and 9-3 summarize the Primary Factor ratings that were generated by the authors of this Manual and which form part of the basis for the ranked sorbent lists in the Guides in Section 1. A specific example is also provided below of the Primary Factor rating process.

Application--

Particulate, pillow, pad, and mat sorbents are candidates for this scenario. The generic sorbents are assigned an optimum application methodology (shovel, throw or blower) relative to small (<500 gals) and medium (500-5000 gals) spills. Not only do most spills fall into these categories but sorbents are unlikely to be used on very large spills, where other countermeasures (e.g., pumping) would usually be more effective. The actual evaluation employed by the authors consisted of determining the relative rate of sorbency application by selecting an optimum application method, estimating a volume rate of application of sorbent based on this method, and calculating the rate required to apply each sorbent based on a given spill size. The sorbency application rate is the rate of sorbent application times the sorption capacity of a sorbent relative to a specific liquid; it is a measure of the rate at which the potential for hazardous liquid sorption is being applied for a particular sorbent. The sorption capacity (experimental, literature, or estimated) for each sorbent compatible with the representative hazardous liquid of interest is used in calculating the sorbency application rate. The application rating assignments are based on the rates obtained. Also, the nominal volume for particulates is estimated at 0.25 cubic feet, while the nominal sorbent unit volume, used for pillows, pads and mats, is the typical volume of the sorbent units in a particular generic class and is based on an assessment of the unit volumes as provided by the manufacturers and given in the respective Sorbent Data Sheet in Section 2. The estimated rates of sorbency aplication for the generic sorbents are compared, whereupon each generic sorbent receives a relative rating as follows:

+ Relatively high rate of sorbency application

0 Moderate rate of sorbency application

- Relatively low rate of sorbency application

TABLE 9-2. LANDSPILL PRIMARY FACTOR PRIORITIZATION - PARTICULATE SORBENTS

Chemical Class	Sorbent Clay	Diatomite	Expanded Mineral	Wood Fiber	Treated Wood Fiber	Poly-propylene	Cross-Linked Polymer	Foamed Glass	Treated Clay/Treated Natural Organic
	A C Av	A C Av	A C Av	A C Av	A C Av	A C Av	A C Av	A C Av	A C Av
1. Acidic compounds, inorganic	+	0	0/+ 0/0 +	+	0	0	0	0/+ 0/0 0	–
2. Acidic compounds, organic	0/0 0/0 +	0/0 0/0 0	0/0 0/+ +	0/0 0/– +	–/0 0/– 0	0/0 0/+ 0	+/+ 0/+ 0	0/0 0/+ 0	–
3. Alcohols & glycols	0/0 0/0 +	0/0 0/0 0	0/0 0/+ +	–/0 –/– +	–/0 0/– 0	–/0 0/+ 0	+/+ 0/0 0	–/0 0/+ 0	–
4. Aldehydes	0/0 0/0 +	0/0 0/0 0	0/0 0/+ +	0/0 0/0 +	–/0 0/– 0	0/0 0/+ 0	+/+ 0/0 0	0/0 0/+ 0	–
5. Aliphatic hydrocarbons	–/0 0/0 +	–/0 0/0 0	–/0 0/+ +	–/0 –/– +	–/– –/– 0	–/0 0/+ 0	+/+ 0/+ 0	–/0 0/+ 0	–
6. Aliphatics, halogenated	–/0 0/0 +	0/0 0/+ 0	–/0 0/+ +	–/0 0/0 +	0/0 0/– 0	–/0 0/+ 0	+/+ 0/+ 0	–/0 0/+ 0	–
7. Amides, anilides & imides	+	0	+	+	0	0	0	0	–
8. Amines, alkyl	0/0 0/0 +	0/0 0/0 0	0/0 0/+ +	–/– –/– +	–/– –/0 0	0/0 0/+ 0	+/+ 0/+ 0	0/0 0/+ 0	–
9. Amines, aryl	0/+ 0/0 +	0/+ 0/0 0	0/0 0/+ +	0/0 0/0 +	–/0 0/0 0	0/+ 0/+ 0	0	0/0 0/+ 0	–
10. Aromatic hydrocarbons	0/0 0/0 +	0/0 0/+ 0	0/0 0/+ +	0/– –/– +	0/– –/– 0	0/0 0/+ 0	+/+ 0/+ 0	0/– 0/+ 0	+/+ 0/+ –
11. Aromatics, halogenated	–/0 0/0 +	0/0 0/+ 0	–/0 0/+ +	–/0 0/0 +	–/0 0/– 0	–/0 0/+ 0	+/+ 0/+ 0	–/0 0/+ 0	–
12. Basic compounds	0/0 0/– +	0	0/– 0/+ +	+	0	0	+/+ 0/0 0	0/– 0/+ 0	–
13. Cyanates & isocyanates	+	0	0/0 0/+ +	0/0 0/0 +	–/0 0/– 0	0/0 0/+ 0	+/+ 0/0 0	0/0 0/+ 0	–
14. Cyanides & nitriles	0/+ 0/0 +	0/+ 0/0 0	–/0 0/+ +	–/0 0/0 +	–/– –/– 0	0/0 0/+ 0	+/+ 0/0 0	0/0 0/+ 0	–
15. Esters & ethers	0/0 0/0 +	0/0 0/0 0	0/0 0/+ +	0/0 0/0 +	–/– 0/– 0	0/0 0/+ 0	+/+ 0/+ 0	–/0 0/+ 0	–
16. Ethers, halogenated	0/0 0/0 +	0/0 0/0 0	0/0 0/+ +	0/0 0/0 +	–/– 0/– 0	0/0 0/+ 0	+/+ 0/+ 0	–/0 0/+ 0	–
17. Halides, inorganic	+	0	0/0 0/+ +	0/0 0/0 +	0	0/0 0/+ 0	+/+ 0/+ 0	0/0 0/+ 0	–
18. Heavy metals	+	0	+	–/0 –/0 +	0	0	0/+ 0/+ 0	0	–
19. Hydrazines & hydrazides	0/0 0/– +	0/0 0/0 0	+	+	0	0/0 0/+ 0	+/+ 0/0 0	0	–
20. Ketones	0/0 0/0 +	0/0 0/0 0	–/0 0/+ +	–/0 –/– +	–/– –/– 0	–/0 0/+ 0	+/+ 0/+ 0	–/0 0/+ 0	–
21. Nitro/nitroso compounds	0/+ 0/0 +	0/0 0/0 0	0/0 0/+ +	0/0 0/0 +	–/0 0/0 0	0/+ 0/+ 0	0	0/0 0/+ 0	–
22. Organophosphates	0/0 0/0 +	0/0 0/0 0	0/0 0/+ +	0/0 0/0 +	–/0 0/– 0	0/0 0/+ 0	+/+ 0/+ 0	0/0 0/+ 0	–
23. Oxides, alkylene	0/0 0/0 +	0/0 0/0 0	–/0 0/+ +	–/0 –/– +	–/– –/– 0	0/0 0/+ 0	+/+ 0/+ 0	–/0 0/+ 0	–
24. Peroxides	0/0 0/0 +	0/0 0/0 0	–/0 0/+ +	0/0 0/0 +	–/0 0/– 0	0/0 0/+ 0	+/+ 0/0 0	–/0 0/+ 0	–
25. Phenols & cresols	0/0 0/0 +	0/0 0/0 0	0/0 0/+ +	0/0 0/0 +	–/0 0/– 0	0/0 0/+ 0	+/+ 0/+ 0	0/0 0/+ 0	–
26. Sulfates & sulfites	0/0 0/0 +	0/0 0/0 0	0/0 0/+ +	–/0 0/0 +	–/0 0/0 0	0/0 0/+ 0	+/+ 0/0 0	–/0 0/+ 0	–
27. Sulfides & mercaptans	0/0 0/0 +	0/0 0/0 0	0/0 0/+ +	0/0 0/0 +	0/0 0/– 0	0/0 0/+ 0	+/+ 0/+ 0	0/0 0/+ 0	–

LEGEND: */* = Small Spill/Medium Spill
A = Application
Av = Availability
C = Collection

+ = Relatively high rating with respect to a particular primary factor
o = Moderate rating with respect to a particular primary factor
– = Relatively low rating with respect to a particular primary factor

TABLE 9-3. LANDSPILL PRIMARY FACTOR PRIORITIZATION - PILLOW, PAD, MAT SORBENTS

	Feathers			Wood Fiber			Treated Wood Fiber			Poly- propylene			Poly- ethylene			Poly- urethane			Cross-Linked Polymer			Foamed Glass		
Chemical Class	A	C	Av	A	C	Av	A	C	Av	A	C	Av	A	C	Av	A	C	Av	A	C	Av	A	C	Av
1. Acidic compounds, inorganic			0			+			0			0			0			0			0	+/0	+/0	0
2. Acidic compounds, organic			0	+/0	0/-	+	0/-	0/-	0	-/-	-/0	0	-/-	-/0	0			0	+/0	0/0	0	+/0	+/+	0
3. Alcohols & glycols			0	0/-	0/-	+	0/-	0/-	0	-/-	-/0	0	-/-	-/0	0	-/-	-/0	0	+/0	0/-	0	0/-	+/+	0
4. Aldehydes			0	+/0	0/-	+	0/-	0/-	0	-/-	-/0	0	-/-	-/0	0	-/-	-/0	0	+/0	0/0	0	+/0	+/+	0
5. Aliphatic hydrocarbons			0	0/-	0/-	+	0/-	0/-	0	-/-	-/0	0	-/-	-/0	0	-/-	-/0	0	+/+	0/0	0	0/-	+/+	0
6. Aliphatics, halogenated			0	0/0	0/-	+	0/-	0/-	0	-/-	-/0	0	-/-	-/0	0	-/-	-/0	0	+/+	0/0	0	0/0	+/+	0
7. Amides, anilides & imides			0			+			0			0			0			0			0			0
8. Amines, alkyl			0	0/-	0/-	+	0/-	0/0	0	-/-	-/0	0	-/-	-/0	0	-/-	-/0	0	+/+	0/0	0	0/-	+/+	0
9. Amines, aryl			0	+/0	0/-	+	0/-	0/-	0	-/-	-/0	0	-/-	-/0	0	-/-	-/0	0			0	+/0	+/+	0
10. Aromatic hydrocarbons	0/-	+/+	0	0/-	0/-	+	0/-	0/-	0	0/-	-/0	0	0/-	-/0	0	0/-	-/0	0	+/0	0/0	0	0/-	+/+	0
11. Aromatics, halogenated			0	0/0	0/-	+	-/-	0/-	0	-/-	-/0	0	-/-	-/0	0	-/-	-/0	0	+/+	0/0	0	0/0	+/+	0
12. Basic compounds			0			+			0			0			0			0	+/0	0/-	0	+/-	+/+	0
13. Cyanates & isocyanates			0	+/0	0/-	+	0/-	0/-	0	-/-	-/0	0	-/-	-/0	0	-/-	-/0	0	+/+	0/0	0	+/0	+/+	0
14. Cyanides & nitriles			0	0/-	0/-	+	0/-	0/-	0	-/-	-/0	0	-/-	-/0	0			0	+/0	0/0	0	+/0	+/+	0
15. Esters & ethers			0	0/-	0/-	+	0/-	0/-	0	-/-	-/0	0	-/-	-/0	0	-/-	-/0	0	+/+	0/0	0	0/-	+/+	0
16. Ethers, halogenated			0	0/-	0/-	+	0/-	0/-	0	-/-	-/0	0	-/-	-/0	0	-/-	-/0	0	+/+	0/0	0	0/-	+/+	0
17. Halides, inorganic			0	0/0	0/-	+			0	0/-	-/0	0	0/-	-/0	0	0/-	-/0	0	+/+	0/0	0	0/0	+/+	0
18. Heavy metals			0	0/0	+/-	+			0			0			0			0	+/+	0/0	0			0
19. Hydrazines & hydrazides			0			+			0	-/-	-/0	0	-/-	-/0	0			0	+/0	0/-	0			0
20. Ketones			0	0/-	0/-	+	0/-	0/-	0	-/-	-/0	0	-/-	-/0	0	-/-	-/0	0	+/+	0/0	0	0/-	+/+	0
21. Nitro/nitroso compounds			0	+/0	0/-	+	0/-	0/-	0	-/-	-/0	0	-/-	-/0	0	-/-	-/0	0			0	+/0	+/+	0
22. Organophosphates			0	+/0	0/-	+	0/-	0/-	0	-/-	-/0	0	-/-	-/0	0	-/-	-/0	0	+/0	0/0	0	+/0	+/+	0
23. Oxides, alkylene			0	0/-	0/-	+	0/-	0/-	0	-/-	-/0	0	-/-	-/0	0	-/-	-/0	0	+/+	0/0	0	0/-	+/+	0
24. Peroxides			0	+/0	0/-	+	0/-	0/-	0	-/-	-/0	0	-/-	-/0	0	-/-	-/0	0	+/0	0/0	0	+/0	+/+	0
25. Phenols & cresols			0	+/0	0/-	+	0/-	0/-	0	-/-	-/0	0	-/-	-/0	0			0	+/+	0/0	0	+/0	+/+	0
26. Sulfates & sulfides			0	+/0	0/-	+	0/-	+/-	0	-/-	-/0	0	-/-	-/0	0	-/-	-/0	0	+/0	0/0	0	+/0	+/+	0
27. Sulfides & mercaptans			0	0/0	0/-	+	0/-	0/-	0	0/-	-/0	0	0/-	-/0	0	0/-	-/0	0	+/+	0/0	0	0/0	+/+	0

LEGEND: */* = Small Spill/Medium Spill
A = Application
Av = Availability
C = Collection

+ = Relatively high rating with respect to a particular primary factor
o = Moderate rating with respect to a particular primary factor
- = Relatively low rating with respect to a particular primary factor

Collection--

The saturated sorbents noted under Primary Factor Evaluation-Application are candidates for this scenario. The generic sorbents are assigned an optimum collection methodology (shovel, pitchfork, or skiploader) relative to the small (<500 gals) and medium (500-5000 gals) spill sizes. An estimated volume rate of collection is then calculated based on the estimated volume for the selected collection method and an estimated rate of collection. Using the appropriate densities for both sorbent and hazardous liquid and a representative sorption capacity (for the respective major generic group of sorbents in which the sorbent is included), the time required to collect the given spill is calculated (see Section 3). Different times for collection based on the given spill size thus allow for discriminating among sorbents as follows:

+ Relatively low time for loaded sorbent collection

0 Moderate time for loaded sorbent collection

\- Relatively high time for loaded sorbent collection

Availability--

Availability is evaluated on a relative scale based on the number of generic sorbent sources. The number of manufacturers, distributors, and suppliers of generic sorbents are tallied based on available data to establish an estimate of sorbent availablity. The rating of availability is as follows:

+ Relatively high availability

0 Moderate availability

\- Relative low availability

Example of Application Factor Evaluation

An example of this evaluation approach as applied to Chemical Class No. 6 is illustrated by the computer printout for the category "Landspill - Small - Application," depicted in Figure 9-1. The sorbency application rate (column H, pounds of hazardous substance sorbed/minute) is the quantity that determines the rating received by a sorbent for the Primary Performance Factor of Application. Since particulate and pillow, pad, and mat sorbents are suitable candidates for landspill applications, both sorbent types are evaluated. The data comprise the assumptions and computations made to obtain the estimated sorbency application rates for the candidate sorbents considered for use with halogenated aliphatics for the stated scenario. Similar assumptions and computations were employed in obtaining the rating assignments for other scenarios.

To obtain the sorbent rating assignments for Chemical Class No. 6

CHEMICAL CLASS: 6

Landspill - Small - Application

Sorbent Class	(A) Nominal Volume (ft3)	(B) Bulk Density (#/ft3)	(C) Sorption Capacity (#/#)	(D) Best App. Method	(E) Movement per min.	(F) Sorbent App. Rate (ft3/m)	(G) Sorbent App. Rate (#/m)	(H) Sorbency App. Rate (#/m)	(I) Rating Assign-ments
PARTICULATES									
Sorbent clay	0.25	34	1.6	Shovel	5	1.25	42.5	68.0	-
Diatomite	0.25	28	2.5	Shovel	5	1.25	35.0	87.5	0
Wood fiber	0.25	12	2.4	Shovel	6	1.5	18.0	43.2	-
Treated WF	0.25	7	2.2	Shovel	6	1.5	10.5	23.1	-
Expanded mineral	0.25	5	6.7	Shovel	6	1.5	7.5	50.25	-
Foamed glass	0.25	2	12.1	Shovel	6	1.5	3.0	36.3	-
CLP	0.25	38	36.9	Shovel	5	1.25	47.5	1752.75	+
TC	0.25	45	-	Shovel	5	1.25	56.25	NA	NA
Polypropylene	0.25	3	11.3	Shovel	6	1.5	4.5	50.85	-
PILLOWS, PADS, MATS									
Wood fiber	0.43	12	2.4	Throw	10	4.3	51.6	123.84	0
Treated WF	0.5	7	2.2	Throw	10	5.0	35.0	77.9	0
Foamed glass	0.5	2	12.1	Throw	10	5.0	10.0	121.0	0
CLP	0.26	38	36.9	Throw	10	2.6	98.8	3645.72	+
Feathers	0.56	0.89	-	Throw	10	5.6	4.984	NA	NA
Polyurethane	0.09	1.4	20.4	Throw	10	0.9	1.26	25.704	-
Polyethylene	0.06	1.5	25.6	Throw	10	0.6	0.9	23.04	-
Polypropylene	0.07	3.0	11.3	Throw	10	0.7	2.1	23.73	-

NOTE: Typical volume of a shovel is estimated at 0.25 ft^3. Sorbency application rate and rating assignments are not available (NA) for sorbents lacking sorption capacity data. The computational series is as follows:

(1) F = E x A, (2) G = F x B, (3) H = G x C.

Figure 9-1. Example of a Computer Printout of Sorbent Rating Assignments

(Figure 9-1), the number of movements of sorbent onto the spill per minute (column E) is estimated by considering the optimum application method (column D), and the bulk density (column B), along with the average worker's ability to perform the prescribed operation. If a typical shovel holds about 0.25 cubic feet of particulate sorbent, the sorbent application rate in cubic feet/minute (column F) is 0.25 times column E. Likewise, the sorbent application rate in cubic feet per minute for a pillow, pad, or mat (column F) is obtained by multiplying the nominal volume of a pillow, pad, or mat (column A) by column E. To obtain the sorbent application rate in lbs/min (column G), column F is multiplied by the sorbent bulk density (column B). In turn, the sorbent application rate (column G) multiplied by the sorption capacity (column C) produces the sorbency application rate (column H). After column H is computed for all representative sorbents that met the Pass/Fail requirements, including data availability, the sorbency application rate values are segregated into three or fewer groups, depending on the distribution. The high, middle, and low group members are assigned a value of +, -, and 0, respectively. The rating assignments, listed in column I, are also found in the "6 Aliphatics, Halogenated" row in Table 9-2, for the particulate sorbents, and in Table 9-3, for the pillow, pad and mat sorbents.

Secondary Factor Evaluation

The narrative that follows explains the rationale and process used to rate Secondary Performance Factors for the Landspill Scenario. Table 9-4 summarizes the Secondary Performance Factor ratings for this scenario.

Reusability—

Evaluation of this factor is based principally on sorbent compressibility. If a loaded sorbent can be easily compressed or squeezed to release the hazardous liquid and thereupon the sorbent returns, more or less, to its original size and shape, the sorbent is considered reusable. More complicated means for releasing the hazardous liquid are considered impractical and infeasible under field conditions. These criteria preclude the reuse of particulate inorganic or mixed sorbents. Other particulate sorbents, although capable of reuse, may not be easily compressed because of their loose configuration. It should be noted that the use of a sorbent that is reusable usually implies reuse or recycling of the hazardous liquid, otherwise the sorbent selected for spill cleanup would also be used for landfill disposal of the hazardous liquid. The rating is as follows:

+ Sorbent can be effectively reused, i.e., is compressible

0 Sorbent can be effectively reused with some difficulty, i.e., is not easily compressible

\- Sorbent cannot be effectively reused, i.e., is relatively incompressible.

Incinerability—

TABLE 9-4. LANDSPILL SECONDARY FACTOR PRIORITIZATION

	Reusability	Incinerability	Wind Resistance	Precipitation Resistance	Sensitive Site Usability	Dense Ground Cover Usability	Rugged Terrain Usability
PARTICULATE SORBENTS							
o Sorbent Clay	-	-	0	-	+	+	+
o Diatomite	-	-	-	-	+	+	+
o Expanded Mineral	-	-	-	-	+	-	+
o Wood Fiber	-	+	-	-	+	-	+
o Treated Wood Fiber	-	+	-	0	+	-	+
o Polypropylene	0	+	-	+	-	-	+
o Cross-Linked Polymer	-	+	-	+	-	+	+
o Foamed Glass	-	0	-	-	+	-	+
o Treated Clay/Treated Natural Organic	-	-	0	+	+	+	+
PILLOW, PAD, MAT SORBENTS							
o Feathers	+	+	0	+	+	-	-
o Wood Fiber	-	+	+	-	+	-	-
o Treated Wood Fiber	0	+	+	0	+	-	-
o Polypropylene	+	+	+	+	+	-	-
o Polyethylene	+	+	0	+	+	-	-
o Polyurethane	+	+	+	+	+	-	-
o Cross-Linked Polymer	-	+	+	+	+	-	-
o Foamed Glass	-	0	+	-	+	-	-

Legend: + = Relatively high rating with respect to a specific secondary factor
0 = Moderate rating with respect to a specific secondary factor
\- = Relatively low rating with respect to a specific secondary factor

For incineration to be a viable option, a sorbent/hazardous liquid pair would have to be combustible. For application of the sorbent selection rationale, however, only the combustibility of the sorbent has been evaluated. Additionally, the assumption is made that any toxic combustion products would be properly scrubbed out by the incinerating system. The rating system used for generic sorbents is as follows:

+ The sorbent is combustible

- The sorbent is not combustible.

Wind Resistance--

Wind can adversely affect the application of certain sorbents. Some particulate sorbents are severely affected by the wind. Also, pillows are less affected than mats or pads. An assessment of the wind effect is made relative to each generic sorbent as follows:

+ Sorbent application is slightly impacted

0 Sorbent application is moderately impacted

- Sorbent application is severely impacted.

Precipitation--

Rain or snow can seriously affect sorbent effectiveness in some cases. If the precipitation does not inordinately increase application/collection difficulties for a specific sorbent and the sorbent exhibits hydrophobic characteristics or has a hazardous liquid/water preference index much greater than unity, the sorbent would be expected to fare well with respect to other sorbents in inclement weather. If precipitation significantly increases the application/collection effort and the sorbent exhibits hydrophilic characteristics, i.e., its hazardous liquid/water preference index is close to or less than unity, the sorbent would be expected to be severely impacted by inclement weather. The rating factor for generic sorbents is as follows:

+ Sorbent effectiveness is not severely limited

- Sorbent effectiveness is severley limited.

Environmental (Sensitive Site) Effects--

Certain sorbents can be used in sensitive areas because they are biodegradable or are a natural component of the environment or can be removed from the environment without significantly disturbing the substrate. Other sorbents are non-biodegradable or disperse in the environment making it difficult to be harvested. The rating factor is as follows:

+ Sorbent may be used in sensitive areas

- Sorbent is not suitable for use in sensitive areas.

Dense Ground Cover--

Certain sorbents can be applied effectively to clean up hazardous liquid spills located in dense ground cover, i.e., heavy foliage, while other sorbents cannot. The latter are usually too light or too bulky and thus will not properly penetrate dense ground cover. Harvesting the sorbents used in this mode has also been considered in the evaluation, although in some cases the ground cover may have to be sacrificed to collect the sorbent. In fact, the spilled hazardous liquid may have already destroyed the ground cover. The rating factor is as follows:

+ Sorbent can be used effectively

- Sorbent cannot be used effectively.

Rugged Terrain--

Not all sorbents can be used effectively in rugged terrain. Although both application and collection are important, this factor was evaluated relative only to application on the assumption that once the spilled hazardous liquid is stabilized (via use of the sorbent) the loaded sorbent can be collected from the rugged terrain by such means as heavy earth moving equipment if necessary. The rating factor discriminates as follows:

+ Sorbent can be used effectively

- Sorbent cannot be used effectively.

SORBENT PERFORMANCE RANKING PROCEDURE - FLOATING SPILL SCENARIO

Pass/Fail Evaluation

Particulate sorbents and heavier-than-water sorbents were eliminated from consideration in the Floating Spill Scenario. Booms were considered only for the flowing mode portion of the scenario. Also, sorbents were not considered for use with a particular chemical class if incompatibilities or non-wetting was noted. Lack of sufficient data was also cause for elimination.

Primary Factor Evaluation

The narrative below explains how the Primary Factor ratings (Application, Collection, Availability, and Hydrophobicity) were developed for the Floating Spill Scenario. Table 9-5 summarizes the Primary Factor ratings that were generated and used as input for the ranked sorbent lists in the Guides in Section 1.

Application--

Pillow, pad, or mat sorbents are viable candidates for the non-flowing

TABLE 9-5. FLOATING SPILL PRIMARY FACTOR PRIORITIZATION

CHEMICAL CLASS	PILLOW, PAD, MAT SORBENTS: Feathers (A C H Av)	Treated Wood Fiber (A C H Av)	Poly-propylene (A C H Av)	Poly-ethylene (A C H Av)	Poly-urethane (A C H Av)	Cross Linked Polymer (A C H Av)	SORBENT BOOMS: Feathers (A C H Av)	Treated Wood Fiber (A C H Av)	Poly-propylene (A C H Av)	Poly-Urethane (A C H Av)
1. Acidic compounds, inorganic	not applicable - water soluble									
2. Acidic compounds, organic	not applicable - water soluble									
3. Alcohols & glycols	not applicable - water soluble									
4. Aldehydes	not applicable - water soluble or non-floating									
5. Aliphatic hydrocarbons	0	0 +/+ - 0	0 -/- 0 0	0 -/- 0 0	0 0/0 0 0	+ +/+ 0 0	0	0 0/0 - 0	+ +/+ 0 0	0 +/+ 0 0
6. Aliphatics, halogenated	not applicable - non-floating (except Allyl chloride and Chloroethane)									
7. Amides, anilides & imides	insufficient data									
8. Amines, alkyl	not applicable - water soluble									
9. Amines, aryl	0	+ +/+ 0 0	0 -/- 0 0	0 -/- 0 0	0 0/0 0 0	0	0	- 0/0 0 0	+ +/+ 0 0	0 +/+ 0 0
10. Aromatic hydrocarbons	0 +/+ 0 0	0 +/0 - 0	0 -/- 0 0	0 -/- 0 0	0 0/- 0 0	+ +/+ 0 0	- +/+ 0 0	- 0/0 - 0	+ +/+ 0 0	0 +/+ 0 0
11. Aromatics, halogenated	not applicable - non-floating									
12. Basic compounds	not applicable - water soluble									
13. Cyanates & isocyanates	not applicable - decomposes or water soluble									
14. Cyanides & nitriles	0	0 +/+ - 0	- 0/0 0 0	- 0/0 0 0	0	0	0	0 0/0 - 0	+ +/+ 0 0	0
15. Esters & ethers	0	0 +/+ - 0	0 -/- 0 0	0 -/- 0 0	0 0/0 0 0	+ +/+ 0 0	0	0 0/0 - 0	+ +/+ 0 0	0 +/+ 0 0
16. Ethers, halogenated	not applicable - non-floating									
17. Halides, inorganic	not applicable - decomposes or water soluble									
18. Heavy metals	not applicable - non-floating									
19. Hydrazines & hydrazides	not applicable - water soluble									
20. Ketones	0	0 +/+ - 0	- -/- 0 0	- -/- 0 0	- 0/0 0 0	+ +/+ 0 0	0	- -/0 - 0	+ +/+ 0 0	0 0/+ 0 0
21. Nitro/nitroso compounds	0	+ +/+ - 0	0 -/- 0 0	0 -/- 0 0	0 0/0 0 0	0	0	0 0/0 - 0	+ +/+ 0 0	0 +/+ 0 0
22. Organophosphates	not applicable - non-floating									
23. Oxides, alkylene	not applicable - water soluble									
24. Peroxides	not applicable - non-floating									
25. Phenols & cresols	not applicable - non-floating									
26. Sulfates & sulfites	not applicable - non-floating									
27. Sulfides & mercaptans	not applicable - non-floating									

LEGEND:

- */* = Small Water Body/Large Water Body
- A = Application
- Av = Availability
- C = Collection
- H = Hydrophobicity
- + = Relatively high rating with respect to a particular primary factor
- 0 = Moderate rating with respect to a particular primary factor
- - = Relatively low rating with respect to a particular primary factor

water mode of this scenario. Only those sorbents that are hydrophobic have been considered. The generic sorbents are assigned an optimum application methodology relative to the size of the water body into which the spill has occurred, regardless of spill size. The relative rates of sorbency application for each optimum method is estimated by a technique similar to that employed for the Landspill scenario and described above.

For flowing water systems, sorbent booms are candidates. For spills into small flowing bodies of water, e.g., creeks, the spill responder may apply the sorbent in an operation that would involve passing a sorbent boom across the waterway. This could be accomplished by casting a tie-line to a team member on the other side of the waterway whereupon the sorbent boom is pulled across. Of course, in a second stage to the operation pillow, pad, or mat sorbents could then be added to the concentrated spill (and collected) in the optimum manner designated for the use of such sorbent in a non-flowing water body.

For spills into large flowing or non-flowing bodies of water, the spill responder may employ a boat or similar conveyence to facilitate the application (and collection) of sorbents. If a means of conveying the sorbent (before deployment) has to be used in its application, the rate of sorbent application is decreased accordingly to reflect the added inconvenience of requisitioning and employing the conveyance.

Each qualifying generic sorbent receives a relative rating as follows:

\+ Relatively high rate of sorbency application

0 Moderate rate of sorbency application

\- Relatively low rate of sorbency application

Collection--

The saturated sorbents cited under <u>Primary Factor Evaluation</u> - Application are candidates for this scenario. An optimum collection method and the weight of loaded sorbent collected per unit time is determined in a fashion similar to that described under the Landspill scenario above. Different times for collection based on a given spill size thus allow for discriminating among sorbents as follows:

\+ Relatively low time for loaded sorbent collection

0 Moderate time for collection

\- Relatively high time for loaded sorbent collection.

Hydrophobicity--

For a sorbent to adequately sorb a free or undissolved hazardous liquid in the presence of water, the sorbent must preferentially sorb the hazardous liquid instead of water, i.e., the sorbent must be hydrophobic. Assessment of the hydrophobic quality of a sorbent is a significant element in determining a sorbent's potential performance in a waterspill application. A measure of the hydrophobic nature of a sorbent is provided via the hazardous liquid/water preference indices contained in Table 2-2.

For the Floating Spill scenario, only those sorbents exhibiting a preference index close to or less than unity are eliminated from consideration for use while those that are not eliminated and have a preference index less than 2 received a "-" rating. Those sorbents exhibiting a preference index greater than 2 should generally be unaffected by the presence of water and thus received a "0" rating. (See Landspill Scenario-Precipitation, this section, for related information.) The rating factor for hydrophobicity involvng a Floating Spill scenario is as follows:

0 Sorbent effectiveness is relatively unaffected by presence of water, i.e., its hazardous liquid/water preference index is 2 or greater.

\- Sorbent effectiveness is significantly limited by presence of water, i.e., its hazardous liquid/water preference index is less than 2.

Secondary Factor Evaluation

The narrative that immediately follows explains the rationale and process used to rate Secondary Performance Factors for the Floating Spill Scenario. Table 9-6 summarizes the Secondary Factor Ratings that were generated for the Floating Spill Scenario. The narrative and rating levels are the same as for the Landspill Secondary Factors, except that only two Factors (Reusability and Wind Scatter Resistance) are included for the Floating Spill. Dense ground cover and rugged terrain play no role in the Floating Spill Scenario while Precipitation Effects are already accounted for under Hydrophobicity. Since the candidate pillow, pad, or mat sorbents and sorbent booms (see Table 9-6) are all combustible and generally recoverable from environmentally sensitive sites (in contrast to particulate sorbents), incinerability and environmental (sensitive site) factors are not included.

Reusability—

Evaluation of this factor is based principally on sorbent compressibility. If a loaded sorbent can be easily compressed or squeezed to release the hazardous liquid and thereupon the sorbent returns, more or less, to its original size and shape, the sorbent is considered reusable. More complicated means for releasing the hazardous liquid are considered impractial and unfeasible under field conditions. These criteria preclude the reuse of particulate inorganic or mixed sorbents. Other particulate sorbents, although capable of reuse, may not be easily compressed because of their loose configuration. It should be noted that the use of a sorbent that is reusable usually implies reuse or recycling of the hazardous liquid, otherwise the sorbent selected for spill cleanup would also be used for landfill disposal of the hazardous liquid. The rating is as follows:

TABLE 9-6. FLOATING SPILL SECONDARY FACTOR PRIORITIZATION

	Reusability	Wind Resistance
NON-FLOWING MODE		
PILLOW, PAD, MAT SORBENTS		
o Feathers	+	0
o Treated Wood Fiber	0	+
o Polypropylene	+	+
o Polyethylene	+	0
o Polyurethane	+	+
o Cross-Linked Polymer	-	+
FLOWING MODE		
SORBENT BOOMS		
o Feathers	+	+
o Treated Wood Fiber	0	+
o Polypropylene	+	+
o Polyurethane	+	+

Legend: + = Relatively high rating with respect to a specific secondary factor
0 = Moderate rating with respect to a specific secondary factor
- = Relatively low rating with respect to a specific secondary factor

+ Sorbent can be effectively reused, i.e., is compressible

0 Sorbent can be effectively reused with some difficulty, i.e., is not easily compressible

- Sorbent cannot be effectively reused, i.e., is relatively incompressible

Wind Resistance--

Wind can adversely affect the application of certain sorbents. Some particulate sorbents are severely affected by the wind. Also, pillows are less affected than mats or pads. An assessment of the wind effect is made relative to each generic sorbent as follows:

+ Sorbent application is slightly impacted

0 Sorbent application is moderately impacted

- Sorbent application is severely impacted.

SORBENT PERFORMANCE RANKING PROCEDURE - LANDFILL SCENARIO

Pass/Fail Performance Factor Evaluation

Sorbents were not considered for chemical classes with which the sorbents were incompatible or non-wettable or for which insufficient data existed. Pillow, pad, or mat sorbents do not appear feasible since they would not exhibit good compaction resistance under landfill pressures and there is no assurance that liquid pockets would not form between these sorbent forms. Also, sorbent booms are impractical for this scenario.

Primary Factor Evaluation

Particulate sorbents are suitable candidates for the scenario involving immobilization of liquid wastes prior to landfilling, in which free liquid in a drum is sorbed to allow for permitted hazardous waste landfill disposal. A reasonable excess of the sorbent would assure that no free liquid would ever be present under shipping or landfill conditions. Although some sorbents may not exhibit good compaction resistance and could release free liquid under landfill pressures, evaluation of the performance of sorbents under pressure was beyond the scope of this project.

Since the Landfill Scenario is not an emergency situation, cost is used as the primary basis for a comparison of the candidate sorbents. Likewise, sorbent availability is not a performance factor in this scenario. For those sorbents considered viable candidates, comparisons are based on the total cost of a specific disposal scenario. Representative sorption capacity (i.e., the sorption capacity of a representative for a sorbent class when contacted with a representative for a chemical class) is used to determine the volume of compatible sorbent that would be required for a given volume of hazardous liquid (see Sections 2 and 3). The Landfill Scenario assumes the following:

1. Total Cost = Cost of Sorbents + Drums + Mixing + Transportation + Disposal
2. The Volume of hazardous liquid spilled is 1000 gallons.
3. The liquid will be soaked up with sorbent and placed in 55-gal drums.
4. Each drum (reconditioned) costs $12 (Ref. 10).
5. The drum volume available for loaded sorbent is 50 gallons (Ref. 10).
6. The mixing cost is $2.00/drum (Ref. 10).
7. The drums wll be transported 100 miles @ $0.05/mi/drum to a permitted landfill site (Ref. 10).
8. The disposal cost is $25.00/drum (Ref. 10).

Based on the assumptions listed above, the Total Cost is computed as follows:

1. lb of hazardous liq = 1000 gal x 8.35 lb/gal x sp gr of liq
2. lb sorbent = lb of hazardous liq/sorption capacity
3. sorbent cost = lb sorbent x cost/lb
4. cu ft of sorbent = lb sorbent/sorbent bulk density (lb/ft^3)
5. gal sorbent = cu ft sorbent x 7.48 gal/cu ft
6. total volume of saturated sorbent = 1000 gal liquid + gal sorbent
7. total drums = total volume of saturated sorbent (gal)/50 (gal/drum)
8. drum cost = total drums x $12/drum
9. mixing cost = total drums x $2/drum
10. transportation cost = total drums x $0.05/mile x 100 mile
11. disposal cost = total drums x $25/drum

The costs for landfill disposal have been calculated for the lowest- and highest-priced sorbent in each generic sorbent class. The results are provided in each Guide in Section 1, under the Landfill Scenario, as a range of costs with the order of listing determined by the lowest cost for each sorbent class.

PROCEDURE FOR COMPUTING AN OVERALL SORBENT PERFORMANCE RATING

For each of the primary and secondary factors, a score (+,0,-) was assigned as described in the previous parts of this section. These primary and secondary factor ratings are given in Tables 9-2 through 9-6 for each sorbent class/chemical class/spill scenario combination (excluding the Landfill Scenario) that is covered in the Guides. For each combination the ratings for the primary factors were doubled, and then the factor ratings were summed to produce one number which is the overall sorbent performance rating for that sorbent class/chemical class/spill scenario combination. The overall sorbent performance ratings were ranked from high to low and the top six were listed under the appropriate Sorbent Selection and Use Guide spill scenario.

As an example, for Chemical Class No. 6 and the Landspill - Small scenario, combining primary factor data from Tables 9-2 and 9-3 and secondary factor data from Table 9-4 produces the following overall values:

PARTICULATES	VALUES
Sorbent Clay	0
Diatomite	-1
Wood Fiber	-1
Treated Wood Fiber	0
Expanded Mineral	-3
Foamed Glass	-4
Cross-Linked Polymer	+3
Polypropylene	-2

PILLOWS, PADS, MATS	VALUES
Wood Fiber	+1
Treated Wood Fiber	+1
Foamed Glass	0
Cross-Linked Polymer	+3
Polyurethane	-1
Polyethylene	-2
Polypropylene	-1

The final overall values are used to compare the sorbents for ranking in the prioritized lists found in the Sorbent Selection and Use Guides. In the example given, the six highest-rated sorbents have been listed for consideration in Sorbent Selection and Use Guide No. 6 under Landspill - Small.

10. References

1. Lyman, Handbook of Chemical Property Estimation Methods, McGraw Hill Publishing Co., Novato, California, 1982.

2. Michalovic, C.K., et.al., Multipurpose Gelling Agent and Its Application to Spilled Hazardous Materials, Calspan Corporation, EPA-600/2-77-151, Cincinnati, Ohio, 1977.

3. Smith, G.F., Performance Testing of Selected Sorbent Booms, Mason & Hanger - Silas Mason Co., Inc., EPA-600/7-78-219, Cincinnati, Ohio, 1978.

4. Robinson, J.S., Ed., Hazardous Chemical Spill Cleanup, Noyes Data Corporation, Park Ridge, New Jersey, 1979.

5. Lafornara, J.P., "Cleanup After Spills of Toxic Substances," J. Water Pollution Control Federation, Vol. 50, No. 4, 1978, pp. 617-627.

6. Shaw, H.S. et.al., Development of A Sorbent Distribution and Recovery System, Seaward International, Inc., EPA-600/7-78-217, Cincinnati, Ohio, 1978.

7. McMahon, L.A., 1983 Dodge Guide to Public Works and Heavy Construction Costs, Annual Edition No. 15, McGraw Hill Information Systems Company, Princeton, New Jersey, 1982.

8. Rental Equipment Register, Rental Equipment Rate Guide for the Year 1983, Miramar Publishing Company, Los Angeles, California, 1983.

9. Gibbons, J.H., Director, Technologies and Management Strategies for Hazardous Waste Control, Office of Technology Assessment, Congress of the United States, Washington, D.C., 1983.

10. Browers, S.D., Understanding Sorbents for Cleaning Up Spills, Plant Engineering, March 18, 1982, pp 219-221.

11. Radecca Lab Notes, Volume 1, No. 2, Radecca, Inc., Austin, Texas, March, 1983.

11. Glossary

ASTM - American Society for Testing and Materials. A scientific and technical organization formed for the development of standards and guidelines.

batch treatment - removal of an insoluble hazardous liquid from water by contact with a sorbent in a mixing vessel for discrete exposure periods.

biodegrade - the ability (of a sorbent) to be broken down and mineralized by microbes, fungi, etc.

CAS number - a 5 to 8 digit identifying number given to each unique chemical substance by the Chemical Abstract Service.

CERCLA - Comprehensive Environmental Response, Compensation and Liability Act - also known as Superfund.

column treatment - removal of an insoluble hazardous liquid from water by continuous flow through a column packed with sorbent material.

cross-linked polymer (CLP) - a plastic sorbent composed of molecules which are lightly cross-linked to each other, imparting imbibing qualities to the plastic.

density, bulk (sorbent) - the weight per unit volume of uncompressed, as-received sorbent.

density, sorbent (no voids) - the weight per unit volume of a sorbent material (i.e., the weight per unit volume of a sorbent without void spaces), viz-a-viz bulk density.

diatomite - a silicious material derived chiefly from the remains of the planktonic algae called diatoms (also called diatomaceous earth).

dust respirator - a respirator designed to provide essentially particulate-free air for breathing.

emulsify - the act of creating a stable dispersion in a two-phase immiscible liquid/liquid system.

expanded mineral - those minerals that expand upon heating to yield low bulk density material, e.g., perlite and vermiculite.

feather sorbent - a sorbent composed of sanitized, dequilled chicken feathers contained in a mesh bag.

foamed glass - amorphous silicate glass foam, consisting of spheroid-shaped particles with numerous cells and characterized by very low bulk density.

GSA - General Services Administration. An agency of the Federal government responsible for the issuance of product qualification standards.

hazardous liquid/water preference index - the ratio of the sorption capacity of a sorbent with a hazardous liquid to the sorption capacity of the same sorbent with water.

hazardous substance - a chemical substance listed in CERCLA Section 101(14).

hydrophilic - that characteristic of a sorbent whereby the sorbent is easily wet by water.

hydrophobic - that characteristic of a sorbent whereby the sorbent is not easily wet by water.

hysteresis - that characteristic of a sorbent that returns it to its original shape after the removal of compressional forces.

incompatible - a condition that describes contact between a sorbent and a hazardous liquid resulting in a reaction that is violent or otherwise hazardous or that results in significant deterioration of the sorbent during the normal period of use.

liquid - for the purposes of this handbook, a substance is a liquid if its melting point is at or below $77^{o}F$ ($25^{o}C$) and its boiling point is at or above $50^{o}F$ ($10^{o}C$).

miscible - the ability of two liquids to dissolve, one in the other, in any proportion.

neat - pure, free from admixture or dilution.

non-wet - a condition wherein a sorbent is unable to soak up and retain a liquid because the liquid does not sufficiently spread on the sorbent surfaces (see Section 8 of the handbook for a discussion of surface tension).

oleophilic - that characteristic of a sorbent whereby the sorbent is easily wet by oil.

overflow dam - a dam that extends to the bottom of a watercourse so that flow is over the top of the dam.

particulate sorbent - finely-divided sorbent.

polyethylene sorbent - a polymer of ethylene that is generally foamed.

polypropylene sorbent - a polymer of propylene where fibers are generally bonded together by heat or needle-punching.

polyurethane sorbent - any of the various polymers that contain the -NHCOO- linkages and are generally foamed.

SCBA - self-contained breathing apparatus. Provides complete respiratory protection in hazardous atmospheres.

skiploader - a specific type of earth moving equipment characterized by a front-mounted scoop with a high degree of maneuverability.

solubility - the degree to which a substance dissolves in water.

sorbency application rate - the rate of sorbent application times the sorption capacity of a sorbent relative to a specific liquid; it is a measure of the rate at which the potential for hazardous liquid sorption is being applied for a particular sorbent.

sorbent - a material that soaks up and retains a liquid by either absorption or adsorption or both.

sorbent blanket - a large, flat sorbent unit usually 24 or 36 inches long, and sometimes as wide, and 1/4 to 1 inch thick.

sorbent boom - a cylindrically configured sorbent usually 8 inches in diameter and 10 feet long and used, connected in series, to clean up floating spills in water bodies.

sorbent clay - a natural earth product composed of fine particles of aluminum silicates and other minerals that is capable of sorbing and holding liquids.

sorbent mat - similar to sorbent blanket

sorbent pad - a small, flat sorbent unit usually 18 inches on a side and 1/4 to 1 inch thick.

sorbent pillow - a sorbent configuration where loose or particulate sorbent is placed in a permeable bag.

sorption capacity - a measure of the degree to which a sorbent can sorb a liquid, expressed in terms of amount of liquid per amount of sorbent.

sparkproof - the ability of a tool to resist sparking during its use. Such tools are generally made out of beryllium-copper alloys.

specific gravity - the ratio of a substance's density to the density of water.

spill, large - a hazardous substance spill of 5000 gallons or more.

spill, medium - a hazardous substance spill of greater than 500 gallons but less than 5000 gallons.

spill, small - a hazardous substance spill of 500 gallons or less.

surface tension - an internal force of a liquid that inhibits a liquid's ability to wet surfaces with which it is in contact.

swelling - an increase in sorbent size due to wetting by a sorbed liquid.

treated clay/treated natural organic sorbent - wood fiber that has been treated with chemicals to increase water repellancy on a substrate of sorbent clay that has been treated with chemicals to increase water repellancy.

treated wood fiber - wood fiber that has been treated with chemicals to increase water repellancy.

ullage - vapor space within a closed vessel containing a liquid and/or solid.

USCG - United States Coast Guard.

viscosity - a measure of the forces that work against movement or flow of liquid when a shearing stress is applied.

wet - a condition wherein a sorbent saturates with a liquid because the liquid spreads on the sorbent surfaces (see Section 8 of the handbook for a discussion of surface tension).

wood fiber sorbent - cellulose-based sorbent derived from the woody portion of plants.

12. Table of Conversions

U.S. Customary Unit			SI	
Name	Abbreviation	Multiplier	Symbol	Name
cubic foot	ft^3	28.32 0.0283	L m^3	litre cubic metre
cubic feet per minute	ft^3/min	0.0283	m^3/min	cubic metres per minute
cubic feet per pound	ft^3/lb	62.4	L/kg	litres per kilogram
cubic feet per second	ft^3/s	28.32	L/s	litres per second
cubic inch	$in.^3$	16.39 0.0164	cm^3 L	cubic centimetres litre
cubic yard	yd^3	0.765 764.6	m^3 L	cubic metre litre
degrees Fahrenheit	^oF	$0.555(^oF-32)$	^oC	degrees Celsius
feet per minute	ft/min	0.00508	m/s	metres per second
feet per second	ft/s	0.305	m/s	metres per second
foot (feet)	ft	0.305	m	metre(s)
gallon(s)	gal	3.785	L	litre(s)
gallons per minute	gal/min	0.0631	L/s	litres per second
inch(es)	in.	2.54	cm	centimetre
inches per hour	in./h	2.54	cm/h	centimetres per hour
mile	mi	1.609	km	kilometre
pound(s)	lb	0.454	kg	kilogram(s)
pounds per cubic foot	lb/ft^3	16.018	kg/m^3	kilograms per cubic metre
pounds per square foot	lb/f^2	4.882×10^{-4} 4.882	kg/cm^2 kg/m^2	kilograms per square centimetre kilograms per square meter
pounds per square inch	$lb/in.^2$	0.0703	kg/cm^2	kilograms per square centimetre
square foot	ft^2	0.0929	m^2	square metre
square inch	$in.^2$	6.452	cm^2	square centimetre

Density of Water = 62.416 lb/ft^3

Other Noyes Publication

MINIMIZING EMPLOYEE EXPOSURE TO TOXIC CHEMICAL RELEASES

by

Ralph W. Plummer, Terrence J. Stobbe, James E. Mogensen, Luanne K. Jeram

West Virginia University

Pollution Technology Review No. 145

This book describes procedures for minimizing employee exposure to toxic chemical releases and suggested personal protective equipment (PPE) to be used in the event of such chemical release. How individuals, employees, supervisors, or companies perceive the risks of chemical exposure (risk meaning both probability of exposure and effect of exposure) determines to a great extent what precautions are taken to avoid risk.

The objective of this study was to obtain information from chemical manufacturers and facilities which use chemicals in other processes, to determine what types of procedures are currently in practice, and to develop a set of recommended procedures and suggested PPE which can be used to minimize the possibility and the risk of toxic chemical releases. In Part I, the authors develop an approach which divides the project into three phases: kinds of procedures currently being used; the types of toxic chemical release accidents and injuries that occur; and, finally, integration of this information into a set of recommended procedures which should decrease the likelihood of a toxic chemical release and, if one does occur, will minimize the exposure and its severity to employees. Part II covers the use of personal protective equipment. It addresses the questions: what personal protective equipment ensembles *are used* in industry in situations where the release of a toxic or dangerous chemical may occur or has occurred; and what personal protective equipment ensembles *should be used* in these situations.

A condensed table of contents listing **part titles, chapter titles and selected subtitles** is given below.

ISBN 0-8155-1131-0 (1987) 257 pages